A2 Revision**Notes**
Biology

• Alan Morris • Margaret Baker •
Series editor: Jayne de Courcy

William Collins' dream of knowledge for all began with the publication of his first book in 1819. A self-educated mill worker, he not only enriched millions of lives, but also founded a flourishing publishing house. Today, staying true to this spirit, Collins books are packed with inspiration, innovation and practical expertise. They place you at the centre of a world of possibility and give you exactly what you need to explore it.

Collins. Do more.

Published by Collins
An imprint of HarperCollins*Publishers*
77–85 Fulham Palace Road
Hammersmith
London
W6 8JB

Browse the complete Collins catalogue at
www.collinseducation.com

© HarperCollins*Publishers* Limited 2006

10 9 8 7 6 5 4 3 2 1

ISBN-13 978 0 00 720686 5
ISBN-10 0 00 720686 0

Alan Morris and Margaret Baker assert the moral right to be identified as the authors of this work

All rights reserved. No part of this publication may be reproduced, stored in a retrieval system, or transmitted in any form or by any means, electronic, mechanical, photocopying, recording or otherwise, without the prior written permission of the Publisher or a licence permitting restricted copying in the United Kingdom issued by the Copyright Licensing Agency Ltd., 90 Tottenham Court Road, London W1T 4LP.

British Library Cataloguing in Publication Data
A Catalogue record for this publication is available from the British Library

Edited by Pat Winter
Production by Katie Butler
Series design by Sally Boothroyd
Illustrations by Kathy Baxendale
Index compiled by Joan Dearnley
Printed and bound by Printing Express, Hong Kong

You might also like to visit
www.harpercollins.co.uk
The book lover's website

CONTENTS

HOW THIS BOOK WILL HELP YOU iv

SECTION I: AQA BIOLOGY A, BIOLOGY B AND HUMAN BIOLOGY

Meiosis	1
Genetics	3
Variation	11
Selection	15
Classification	17
Ecology	19
Biochemistry of photosynthesis	27
Biochemistry of respiration	29
Homeostasis	35
Nervous system	39
The eye	45

SECTION II: AQA BIOLOGY A

Water transport in plants	49
The liver and the kidney	53
Gas exchange	59
Transport of gases	63
Digestion	67
Metamorphosis and diet	73
Behaviour	75

SECTION III: AQA BIOLOGY B

The kidney	77
The brain	83
Muscles	85
Microbes and disease	89

SECTION IV: AQA HUMAN BIOLOGY

Reproduction	105
Growth and development	111
Digestion	113
Dietary requirements	119
Transport of gases	123
Muscles	127
Senescence (ageing)	131

INDEX 132

HOW THIS BOOK WILL HELP YOU

We have planned this book to make your revision as easy and effective as possible.

Here's how:

SHORT, ACCESSIBLE NOTES THAT YOU CAN INTEGRATE INTO YOUR REVISION FILE

Collins Revision Notes A2 Biology Notes have been prepared by top examiners who know exactly what you need to revise in order to achieve a top grade.

You can *either* revise entirely from this book *or* you can tear off the notes and integrate them into your own revision file. This will ensure that you have the best possible notes to revise from.

STUDENT-FRIENDLY PRESENTATION

The notes use lots of visual aids – diagrams, tables, charts etc. – so the content is easier to remember.

There is also systematic use of colour to help you revise:

> **MUST REMEMBER**
> Red panels stress ideas that you must express clearly in order to gain full marks.

> **MUST TAKE CARE**
> Purple panels highlight particularly tricky areas.

Red type identifies key biology terms.
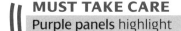 is used for key definitions.
Yellow highlight is used to emphasise important words and phrases

CONTENT MATCHED TO YOUR SPECIFICATION

This book covers the A2 course for three specifications:
AQA Biology Specification A
AQA Biology Specification B
AQA Human Biology

You need to revise the whole of Section I whichever of these specifications you are following. Section II contains additional topics for AQA Biology Specification A. Section III contains additional topics for AQA Biology Specification B. Section IV contains additional topics for AQA Human Biology.

GUIDANCE ON EXAM TECHNIQUE

This book concentrates on providing you with the best possible revision notes.

Knowing the facts is vital – but you may also want help with answering exam questions. That is why we have also produced an Exam Practice book which you can use alongside these Revision Notes: *Collins Do Brilliantly A2 Biology*.

Using both these books will ensure that you achieve the highest possible grade in your A2 Biology exams.

MONOHYBRID CODOMINANT CROSS

In a **monohybrid codominant cross**, only **one gene** is considered.

EXAMPLE: HAIR COLOUR IN SOME BREEDS OF CATTLE
- There are **two alleles** controlling hair colour, and both are expressed:
  They are represented with a letter for the gene for colour **C**, with the superscripts **R** or **W** to represent the specific allele.
  C^R = red hair
  C^W = white hair

Genotypes	Phenotypes
$C^R C^R$	Red hair
$C^W C^W$	White hair
$C^R C^W$	Roan hair (mixture of red and white)

- Must note that there are now **three phenotypes** and not just two.

A MONOHYBRID CROSS BETWEEN TWO HETEROZYGOUS INDIVIDUALS

The **genotype** of both is $C^R C^W$.

Parental phenotypes	Roan cattle × Roan cattle
Parental genotypes	$C^R C^W$ × $C^R C^W$
Gametes	C^R or C^W C^R or C^W
Punnett square	

Gametes	C^R	C^W
C^R	$C^R C^R$	$C^R C^W$
C^W	$C^W C^R$	$C^W C^W$

Offspring genotypes	$C^R C^R$ $C^R C^W$ $C^W C^R$ $C^W C^W$
Offspring phenotypes	Red Roan Roan White

- $\frac{1}{4}$ of the offspring are red ($C^R C^R$)
- $\frac{1}{2}$ of the offspring are roan ($C^R C^W$)
- $\frac{1}{4}$ of the offspring are white ($C^W C^W$)

Result: The cross produces the **monohybrid codominant ratio of 1 : 2 : 1**.

> **MUST TAKE CARE**
> $C^R C^W$ and $C^W C^R$ both result in the same phenotype, roan. It does not matter in which order they are written.

MONOHYBRID MULTIPLE ALLELE CROSS

In a **monohybrid multiple allele cross**, only one gene is considered.

EXAMPLE: BLOOD GROUPS IN HUMANS
- There are three different alleles controlling four different blood groups:
  The gene is represented by the letter I, and the superscripts A, B and o represent the specific alleles.
  I^A and I^B are both dominant to I^o.
  I^A and I^B are codominant.

Genotype	Blood group
$I^A I^A$ or $I^A I^o$	A
$I^B I^B$ or $I^B I^o$	B
$I^A I^B$	AB
$I^o I^o$	o

> **MUST REMEMBER**
> To work out the genotype of blood groups A or B, must consider the offspring produced when they mate. Example: If a group A individual mates with a group B individual, and produce an offspring with blood group o, they must each have contributed an I^o allele and thus they must be $I^A I^o$ and $I^B I^o$.

SEX LINKAGE

A **phenotypic sex-linked characteristic** is produced by a gene on the X chromosome.

Male	Female
Sex chromosomes: **XY**	**XX**
gene A on X chromosome; no allele on the Y chromosome	two alleles of gene A
• The X chromosome is of medium length. • The Y chromosome is small. • There may be no allele on the Y chromosome to correspond to an allele on the X chromosome. • Therefore, since only one allele of the gene is present, one recessive allele can be expressed.	Both alleles need to be recessive before they are expressed.

This type of inheritance is more noticeable if the phenotype is harmful or causes a problem.

Haemophilia and colour blindness are examples of conditions with sex-linked genes.

MUST REMEMBER

- Although sex-linked conditions tend to be more frequent in males, females can also show them, but only when **two recessive alleles are present**.
- In other animals, the chromosomes that determine sex may be reversed, e.g. males are XX and females XY – must read a question carefully to check.

For practice in answering A2 Biology questions, why not use *Collins Do Brilliantly A2 Biology*?

DIHYBRID DOMINANT/RECESSIVE CROSS

In a **dihybrid dominant/recessive cross**, **two genes** are considered.

EXAMPLE: GENE FOR FLOWER COLOUR AND GENE FOR SEED SHAPE
- There are **two different alleles** of both genes:
  P = purple flower
  p = white flower
  W = wrinkled seed
  w = smooth seed

A DIHYBRID CROSS BETWEEN TWO HETEROZYGOUS INDIVIDUALS

The **genotype** of both is PpWw.

Parental phenotypes	Purple flower × Purple flower
	Wrinkled seed × Wrinkled seed
Parental genotypes	PpWw × PpWw
Gametes	(PW) or (Pw) or (pW) or (pw) (PW) or (Pw) or (pW) or (pw)

Offspring genotypes

Gametes	PW	Pw	pW	pw
PW	PPWW	PPWw	PpWW	PpWw
Pw	PPWw	PPww	PpWw	Ppww
pW	PpWW	PpWw	ppWW	ppWw
pw	PpWw	Ppww	ppWw	ppww

Offspring phenotypes

- $\frac{9}{16}$ of the offspring show both dominant phenotypes
  purple flower and wrinkled seed: (P_W_)

- $\frac{3}{16}$ of the offspring show one dominant and one recessive phenotype
  purple flower and smooth seed **or**
  white flower and wrinkled seed:
    P_ww or ppW_

- $\frac{1}{16}$ of the offspring show both recessive phenotype
  white flower and smooth seed:
    ppww

Result: The cross produces the **dihybrid ratio of 9 : 3 : 3 : 1**.

CHI-SQUARED TEST

- The **chi-squared test** is a statistical test.
- It can be used to find the **probability** (or likelihood) of the results of a cross being significantly close to **predicted** (or expected) ratios.

EXAMPLE
If two heterozygous purple plants are crossed (see monohybrid cross, page 4) and the results recorded, we would expect a 3 : 1 ratio.

Purple	White	Ratio
300	100	obviously 3 : 1
20	380	obviously not 3 : 1
258	142	not obvious whether it is significantly close to 3 : 1

With any statistical test we need to write a **null hypothesis**:

- There is no difference between the observed values and the expected values, and any difference is due to chance.

Chi squared equation

$$\chi^2 = \sum \frac{(O-E)^2}{E}$$

O = number actually observed or counted
E = number expected if the ratio were correct
Σ = 'the sum of'

In this example, the observed numbers are:
Purple = 258
White = 142

- The expected numbers have to be calculated.

WORKING OUT THE EXPECTED NUMBERS IF THERE IS A 3 : 1 RATIO

Add all the observed numbers together	The ratio expected is 3 : 1	Divide 400 by 4	Then multiply 3 × 100	Then multiply 1 × 100
258 + 142 = 400	3 + 1 = 4	400/4 = 100	300	100

Therefore the expected numbers are:
  Purple 300
  White 100

- These figures are now put into the equation.
- To make it easier, put in a table.

Phenotype	Observed number, O	Expected number, E	O – E	(O – E)²	$\frac{(O-E)^2}{E}$
Purple	258	300	–42	1764	1764/300 = 5.88
White	142	100	+42	1764	1764/100 = 17.64

$$\chi^2 = \sum \frac{(O-E)^2}{E} = 23.52$$

- The chi squared value (χ^2) is looked up in a **table of probability**.
- The table has a number of rows representing **degrees of freedom**.
- The number of degrees of freedom for a particular set of results is:
  number of categories minus one
- The number of categories = **number of different phenotypes**.
  – There are 2 phenotypes: purple and white.
  – Therefore degrees of freedom = 2 – 1 = 1.

The chi-squared value is looked up on the line which represents the correct degree of freedom, in this case, 1.

SIMPLIFIED TABLE OF PROBABILITY

Degrees of freedom	Probability, p					
	0.9	0.5	0.3	0.1	0.05	0.01
1	0.016	0.46	1.07	2.71	**3.84**	6.64
2	0.21	1.39	2.41	4.61	5.99	9.21
3	0.58	2.37	3.67	6.25	7.82	11.34

Orange = the row to look at: it is the row for 1 degree of freedom.
Turquoise = the probability (0.05) to consider as the **significant value**.
Blue = the critical value with which to compare the **chi-squared value**.

- The calculated value of χ^2 is 23.2 which is greater than the critical value of 3.84.
- Therefore the probability value is less than 0.05 ($p < 0.05$).

WHAT DOES THIS MEAN?

It means:
- that the null hypothesis is rejected.
- that the ratio of purple to white flowers of 258 : 142 is not likely to be a 3 : 1 ratio.

MUST TAKE CARE

Must never write that the null hypothesis is 'right' or 'wrong'. Instead, statistical tests only allow us to either **accept** or **reject** the hypothesis.

SIMPLIFYING THE FINDINGS

If the probability is greater than 0.05 (to the left of 0.05)	If the probability is less than 0.05 (to the right of 0.05)
• The observed figures are significantly close to the expected figures, • and we must **accept** the null hypothesis. • Any difference is due to chance.	• The observed figures are **significantly different** from the expected figures, • and we must **reject** the null hypothesis. • Any difference is **not** due to **chance**.

VARIATION

Members of the same species have common characteristics. But those characteristics are not identical: they show **variation**.

- Variation is caused by an interaction between the **environment** and the **genotype**.
- Thus, the resulting **phenotype** of an individual is due to both genotype and environment.

CAUSES OF VARIATION

The causes of variation in **genotype** are:

- crossing over and independent assortment in meiosis
- gene mutation
- recombination
- polygenic inheritance.

Environmental factors also cause variation in **phenotype**.

Causes of variation	Explanation
Crossing over (See 'Meiosis', page 2.)	When homologous chromosomes come together, alleles can be exchanged.
Independent assortment (See 'Meiosis', page 2.)	There is a random combination of maternal and paternal chromosomes in each gamete.
Gene mutation Gene mutations include: • Substitution • Addition • Deletion	Changes occur in the DNA, at the level of the chromosome or of the gene. Original base sequence AAA|TTT|CCC|GGG → AAA|CTT|CCC|GGG C substituted for T → AAA|CTT|TCC|CGG|G C added → AAA|TTC|CCG|GG T deleted
Recombination	The fusion of two gametes (egg and sperm) produces a new mix of genetic material.
Polygenic inheritance	This is when two or more genes control many phenotypic characteristics. **Example:** When a characteristic is controlled by two genes, each with two alleles, it is possible for nine different genotypes to control the phenotype.
Environmental factors • Nutrients • Light intensity • Oxygen concentration • Carbon dioxide concentration • Temperature	These factors have an effect on the phenotype created by the genotype.

TYPES OF VARIATION

Discontinuous variation	Continuous variation
Discrete phenotypic characters • Tongue rolling – either can or cannot roll tongue • Seed mass – a range of heavy seeds and of light seeds **There is always a gap in the data.**	**Non-discrete phenotypic characters** • Height – examples of all intermediate heights **There is never a gap in the data.**
• Discontinuous variation is controlled by a single gene. • The gene has two or more alleles.	• Continuous variation is controlled by a number of genes. • This means that the characteristic shows **polygenic inheritance**.

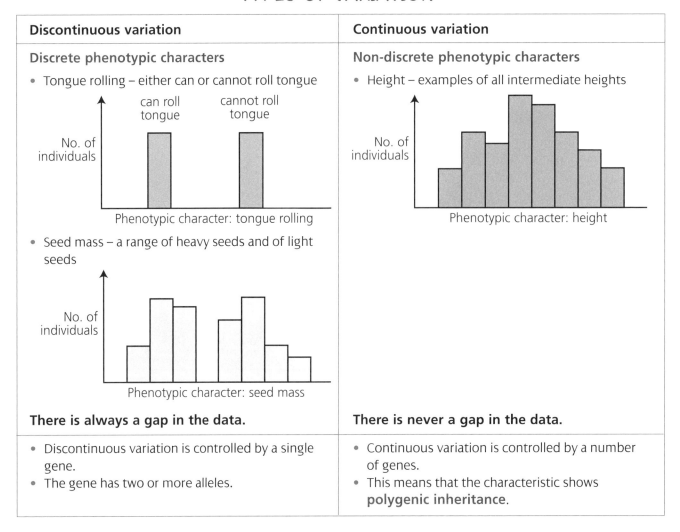

HARDY–WEINBERG PRINCIPLE – ALLELE FREQUENCIES IN A POPULATION

The **Hardy–Weinberg principle** predicts that in a population the **frequency** or **proportion** of each allele of a gene will remain the same from generation to generation. This happens even though the actual number of each allele in the population changes. This is true provided that:

- there is a large population,
- there is random mating (i.e. no selection),
- all genotypes produce individuals that are equally fertile,
- there is no mutation,
- there is no migration into or out of the population.

The Hardy–Weinberg principle involves two concepts:

1. The **gene pool**. It is the total number of alleles of a single gene that are present in a population at a given time.
   Example: In a population of 10 000 individuals, each with a genotype of AA, Aa or aa, each member of the population has 2 alleles. Therefore, the population has a gene pool of 20 000 alleles.

2. **Allele frequency**. It is the proportion of each allele in the gene pool.
   Example: 15 000 A alleles in the above gene pool gives a frequency for the A allele of 15 000/20 000 = 0.75.

MUST REMEMBER

Frequencies are always expressed as **decimals**.

FURTHER EXPLANATION

EXAMPLE A: EXPLAINING GENE POOL AND ALLELE FREQUENCY
- All the mice in a population of 1000 are brown and have the genotype Bb.
- Therefore there are 1000 B alleles and 1000 b alleles in the gene pool of 2000 alleles.
- Thus the frequency of the B allele is 0.5 and the frequency of the b allele is 0.5.

EXAMPLE B: WORKING OUT FREQUENCIES OF GENOTYPES

The Hardy–Weinberg principle involves the frequencies of alleles and the frequencies of genotypes in a given population.

- Let's say a gene has two alleles, a dominant A and a recessive a.
- Three genotypes are possible for this gene: AA, Aa and aa, and are present in the individuals of a population.
- Symbols are used to represent the frequencies of these alleles in the gene pool:
  p = frequency of the allele A
  q = frequency of the allele a

Since there are only two alleles of this gene, the sum of the two frequencies must have a value of 1:

$$p + q = 1$$

However:
- We cannot tell which alleles of a gene an organism possesses.
- What we see is its phenotype.
- The **Hardy–Weinberg equation** enables us to calculate allele frequencies and genotype frequencies by observing the frequencies of phenotypes.

We can represent the frequency of genotypes in a population:
- The frequency of the genotype AA = p^2
- The frequency of the genotype Aa = $2pq$
- The frequency of the genotype aa = q^2

Since these are only genotypes of this gene, the sum of the three frequencies must have a value of 1:

$$p^2 + 2pq + q^2 = 1$$

SUMMARY

Expression	Explanation
• $p + q = 1$ This shows the frequency of the **alleles** in a population.	p = frequency of dominant allele A q = frequency of recessive allele a
• $p^2 + 2pq + q^2 = 1$ This shows the frequency of **genotypes** in a population.	p^2 = frequency of homozygous dominant genotype (AA) q^2 = frequency of homozygous recessive genotype (aa) $2pq$ = frequency of the heterozygous genotype (Aa)

Given any one of the values for p, q, p^2 or q^2, it is possible to calculate all the others.

MUST REMEMBER

- Do not need to know how the formula $p^2 + 2pq + q^2 = 1$ is derived.
- However, must learn how to use it.
- The only **phenotype** which has a **genotype** that is known is the **homozygous recessive**.
- Must identify a homozygous recessive frequency value in a question, and know that it is the value of q^2.

MUST TAKE CARE

- Must always carry out the calculations using **decimals**.
- If frequencies are given in percentages, must always change to decimals before using the equation, e.g. change 64% to 0.64.

WORKED EXAMPLE

A dominant allele A has a frequency of 0.6 in a population. Calculate the frequency of the heterozygous genotype in the population.

Frequency of dominant allele = p
$p = 0.6$
$p + q = 1$
$q = 1 - p$
$\quad = 1 - 0.6$
$q = 0.4$

Frequency of heterozygous genotype = $2pq$
$\quad = 2 \times 0.6 \times 0.4$
$\quad = 0.48$

WORKED EXAMPLE

In a population, 64% of people can taste PTC. The ability to taste PTC is determined by a dominant allele. Calculate the frequency of the dominant allele.

- PTC tasters include both dominant homozygous and heterozygous genotypes.
- The recessive homozygotes make up the rest of the population:
  $100 - 64 = 36\%$
- The frequency of the recessive homozygotes is 36% or 0.36:
  Therefore $q^2 = 0.36$
  $q = \sqrt{0.36}$
  $q = 0.6$
- But $p + q = 1$
  So frequency (p) of the dominant allele:
  $p = 1 - q$
  $\quad = 1 - 0.6$
  $p = 0.4$ or 40%

SELECTION

Evolution can be described as:

changes which occur in the characteristics of a population over many generations, allowing the population to adapt to its environment.

This may result in the formation of a new species. This process is known as **speciation**.

Charles Darwin first developed the idea of **natural selection** as the mechanism for evolutionary change. This mechanism can be summarised as follows:

- The ability to survive depends on the specific characteristics of an organism.
- All characteristics vary within a species.
  - Some variations are advantageous and some are disadvantageous.
- Populations produce more offspring than can survive.
  - If offspring receive advantageous characteristics from their parents, their chances of survival are increased.
- The survivors mature, reproduce and pass on the characteristics.
  - In this way, the **frequency** of the **advantageous allele** within a population will increase.
- This leads to a change in the characteristics of a species over time.

> **MUST REMEMBER**
>
> Selection is a process in which the best adapted organisms in a population survive and reproduce, passing on their alleles to the next generation.

There are many examples of selection, but the five underlying principles are the same in each case.

Underlying principle	Example: The evolution of bacteria resistant to antibiotic X
1. The living organisms in a population **vary**.	In a population of bacteria, some will be resistant to antibiotic X while some will not.
2. Some of this variation is **genetic** and originally arose as a result of **mutation**.	The resistance to antibiotics is genetic – controlled by genes. Different alleles of these genes originally arose as a result of mutation.
3. Selection operates on this population. Some varieties have an **advantage** and some have a **disadvantage**.	When a patient is treated with antibiotic X, only those bacteria which are resistant will survive. The presence of an antibiotic is thus the **selection pressure**.
4. The varieties that have an advantage, **survive** and **reproduce**.	The bacteria with the favourable alleles have an advantage. They survive and reproduce.
5. They pass on the favourable alleles to the next generation, ultimately leading to a **change** in the overall **genetic makeup** of the population.	In reproduction, these alleles are passed on to the next generation. Eventually this leads to a population of bacteria with a **greater frequency of alleles** for antibiotic resistance.

THREE TYPES OF SELECTION

- We look at a particular characteristic within a population that shows **continuous variation** between one extreme and the other.
- For such a characteristic, selection can operate in three different ways.

Stabilising selection
- Selection operates against both extremes.
- The numbers of individuals at the two extremes fall.
- The numbers rise over the middle range.

Directional selection
- Selection operates against one extreme.
- The numbers of individuals at that extreme fall.
- The numbers rise at the other extreme.

Disruptive selection
- Selection against the mean.
- The numbers of individuals around the mean fall.
- The numbers rise of both other extremes.
- Two different populations gradually evolve.

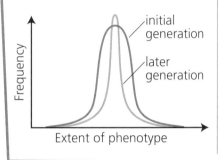
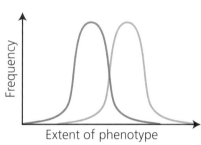
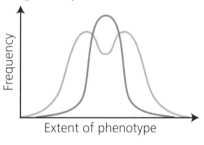

SPECIATION – FORMING A NEW SPECIES

Definition of a **species**:

> A species is a group of interbreeding organisms that produce viable and fertile offspring.

> **MUST TAKE CARE**
> The converse of the definition of a species is also true:
> - If two groups cannot interbreed to produce fertile and viable offspring, they must be **different** species.

- As long as two populations are able to interbreed, they are unlikely to evolve into distinct species.
- They must somehow undergo a period of **reproductive isolation**.
- During this period, genetic differences between the populations increase, through **mutation** and **selection** under different environmental conditions.
- Eventually, the two populations are unable to interbreed and have become distinct species.

There are several ways in which populations can be reproductively isolated:
- Physical – there may be a sea, mountain or desert between them.
- Behavioural – populations may have different courtship patterns.
- Temporal (time based) – plants may flower at different times.

TWO MECHANISMS OF SPECIATION

Allopatric speciation	Sympatric speciation
• **A physical barrier** separates a population so that it occupies two areas. • Geographical isolation occurs. • The environmental conditions or selection pressures are different in the two areas. • Different phenotypes are favoured in the two areas. • Natural selection occurs. • Different alleles are passed on. • The original **phenotype** changes. • If the separated organisms meet they will not be able to breed. • A new species has formed.	• **No physical barrier** separates the population. • Members of the population develop different niches. • They have different food sources or habitats. • **Disruptive selection** occurs. • Two extremes of the population develop. • **Assortative mating** takes place (each type prefers mates like itself). • A new species has formed.

CLASSIFICATION

GROUPING ORGANISMS

- The **classification system** puts similar organisms into groups.
- Each major group contains a number of smaller groups.
- Each smaller group contains a number of smaller groups, and so on.
- Organisms in smaller groups have more characteristics in common than organisms in larger groups.

CLASSIFICATION GROUPS

Name of group	Groups for humans	Groups for lions
Kingdom	Animalia	Animalia
Phylum (sub-phylum)	Chordata Vertebrata	Chordata Vertebrata
Class	Mammalia	Mammalia
Order	Primata	Carnivora
Family	Hominidae	Felidae
Genus	*Homo*	*Panthera*
Species	*sapiens*	*leo*

MUST REMEMBER

Must make up a mnemonic for the names of the groups and their order, biggest to smallest, such as: Kim Plans Competing On Four Game Shows.

The biological name of each species is derived from its genus and species names:

- The genus starts with a capital letter.
- The species starts with a small letter.
- The full name is always printed in italics, e.g. *Homo sapiens* and *Panthera leo*.

For practice in answering A2 Biology questions, why not use *Collins Do Brilliantly A2 Biology*?

THE FIVE-KINGDOM CLASSIFICATION

All living organisms are classified into five kingdoms.

Kingdom	Example	Characteristics
Prokaryotae	Bacterium (capsule, murein cell wall, loop of DNA, cell surface membrane; 0.001 mm)	**Single-celled** Prokaryotic cell structure: • No nucleus or membrane-bound organelles • Cell wall made of murein • **Nutrition**: autotrophic (photosynthesis or chemosynthesis)
Protoctista	Amoeba (nucleus, contractile vacuole; 0.1 mm) — single celled Alga (e.g. seaweed) — Kelp (a multicellular seaweed); 1 m	**Single-celled or multicellular** Eukaryotic cell structure: • Cell wall (when present) of polysaccharide • **Nutrition**: some are autotrophic, some heterotrophic, some both
Fungi	Yeast (nucleus, yeast cell; 0.1 mm) Mushroom (fruiting body made of hyphae, spores)	**Mostly multicellular but some single-celled** Eukaryotic cell structure: • Cell wall of chitin • Most consist of thread-like filaments – hyphae • Produce spores for dispersal • **Nutrition**: heterotrophic
Plantae	Moss (1 cm), Conifer, Fern, Flowering plant	**Always multicellular** Eukaryotic cell structure: • Cellulose cell wall • Many cells possess chloroplasts • Complex life cycles: alternate between a sexually reproducing haploid generation and an asexually reproducing diploid generation • **Nutrition**: autotrophic (photosynthesis)
Animalia	Earthworm, Human	**Always multicellular** Eukaryotic cell structure: • No cell wall • Have a nervous system • **Nutrition**: heterotrophic, with a digestive cavity

ECOLOGY

DEFINITIONS OF ECOLOGICAL TERMS

Population	All the organisms of the same species found in a particular area, present at the same time
Community	All the populations found in a particular area, present at the same time
Environment	The living organisms (biotic factors) and also the non-living conditions (abiotic factors) which surround a group of organisms
Ecosystem	The interaction between the living (biotic) and non-living (abiotic) parts of the environment
Habitat	Where an organism lives – its address
Niche	Where an organism is found, and also what it does there

DISTRIBUTION

- A living organism will exist in a particular area if it can tolerate the **abiotic** factors.
- The numbers found of that species in a particular area will be influenced by **biotic** factors.

SAMPLING

We sample because it would be too time consuming to record everything present in an ecosystem. But the sample must be:

- large enough to be **representative** of the whole population.
- random, to prevent **bias**.
- systematic, to prevent bias.

RANDOM SAMPLING

- Use tapes placed at right angles to produce a **grid**.
- Use a **random numbers table** to generate pairs of numbers to be used as **co-ordinates** for areas on the grid.
- Sample in those areas using a **quadrat**.

SYSTEMATIC SAMPLING

- Use a tape laid across the ecosystem – a **transect**.
- Sample through all the areas which show differences.
- Take samples at regular intervals along the transect, e.g. every 5 metres.

SAMPLING PLANTS AND OTHER NON-MOVING ORGANISMS

A quadrat is suitable for stationary organisms, and can be used to measure:

- population density, e.g. number of organism per m^2.
- frequency, e.g. presence or absence of organism per 100 quadrats.
- percentage cover, e.g. area of the quadrat taken up by the organism.

SAMPLING MOVING ORGANISMS

To sample moving organisms, e.g. mice:

- the mark-release-recapture method is used.
- numbers can be estimated using the **Lincoln Index**:

$$\text{Population size} = \frac{\text{Total number of organisms marked and released} \times \text{Total number of organisms captured in 2nd sample}}{\text{Number of marked organisms in 2nd sample}}$$

The **Lincoln Index** involves these assumptions:

- Released animals mix thoroughly in the population.
- Marking does not affect the animals.
- No migration occurs into or out of the population.
- There are no births or deaths within the population.

WORKED EXAMPLE

The mark-release-recapture method is used on a population of beetles. Calculate the population size from the following:

Number of beetles caught, marked and released = 46
Number of beetles caught in the second sample = 40
Number of marked beetles in the second sample = 16

$$\text{Population size} = \frac{\text{Total number of organisms marked and released} \times \text{Total number of organisms captured in 2nd sample}}{\text{Number of marked organisms in 2nd sample}}$$

$$\text{Population size} = \frac{46 \times 40}{16} = \frac{1840}{16}$$

Estimated population size = 115

> **MUST TAKE CARE**
> - Must be careful to identify the total number in the second sample correctly. For example, 'The second sample contained 17 marked and 41 unmarked individuals.'
> - Must not forget to **add** them together (58) to get the **total** number in the second sample.

DIVERSITY

Diversity describes a community in terms of the number of species and the number of individual organisms present.

The **index of diversity**:
- is a numerical value that can be calculated to represent diversity.
- is used to compare diversity in two areas.

> **EQUATION TO CALCULATE THE INDEX OF DIVERSITY**
>
> The equation is:
>
> $$d = \frac{N(N-1)}{\Sigma n(n-1)}$$
>
> where:
> d = index of diversity
> N = total number of organisms of all species
> n = total number of organisms of each species
> Σ = the sum of

WORKED EXAMPLES

Example 1: Sand dune – closest to the sea

Organisms present	Number
Marram grass	20
Lyme grass	2

$$d = \frac{N(N-1)}{\Sigma n(n-1)}$$

N = total number of species
   = 22
n = number of organisms of each species
   = 20 and 2

Put these values in the equation:

$$= \frac{22(22-1)}{20(20-1) + 2(2-1)}$$

$$= \frac{22 \times 21}{20 \times 19 + 2 \times 1}$$

$$= \frac{462}{380 + 2}$$

$$= \frac{462}{382}$$

$d = 1.2$

In this harsher environment:
- **fewer** species are present and the index of diversity is lower.
- **abiotic** factors are important in determining **which species are present**.

Example 2: Sand dune – furthest from the sea

Organisms present	Number
Calluna	10
Gorse	6
Bell heather	12
Cross-leaved heath	2

$$d = \frac{N(N-1)}{\Sigma n(n-1)}$$

N = total number of species
   = 30
n = number of organisms of each species
   = 10 and 6 and 12 and 2

Put these values in the equation:

$$= \frac{30(30-1)}{10(10-1) + 6(6-1) + 12(12-1) + 2(2-1)}$$

$$= \frac{30 \times 29}{10 \times 9 + 6 \times 5 + 12 \times 11 + 2 \times 1}$$

$$= \frac{870}{19 + 30 + 132 + 2}$$

$$= \frac{870}{183}$$

$d = 4.7$

In this less harsh environment:
- **more** species are present and the index of diversity is higher.
- **biotic factors** are important in determining the **number** of individuals in each species that survive.

SUCCESSION

When other organisms gradually replace organisms in a community:
- **Pioneer** plants that can tolerate harsh environments grow.
- Pioneer plants change the environment, making it less hostile, for example:
  - soil is stabilised
  - pH of soil is changed
  - water in soil is retained
  - humus is increased.
- These changes produce conditions that enable less tolerant plant species to survive.
- The new species now grow and out-compete some of the existing species which die.
- This is repeated again and again.
- Each stage in the succession is called a **sere**.
- The end of the succession produces a stable ecosystem – the **climax community**.

Examples of succession
- Sand dunes become a forest.
- Lakes silt up and become dry land.

SUCCESSION IN DIFFERENT ENVIRONMENTS

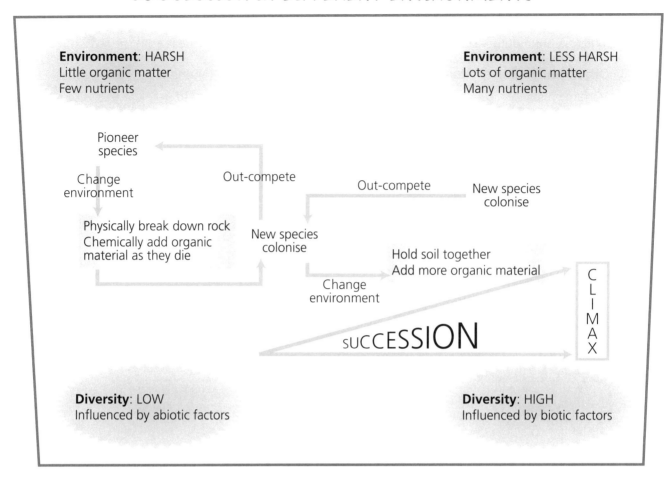

TROPHIC LEVELS AND ENERGY TRANSFER

- The **trophic level** refers to what the organism feeds on.
- **Energy** is transferred – passed on – from one trophic level to the next.

Trophic level	What it 'feeds' on	Detail
Producer	**Produces** organic molecules such as carbohydrates which contain energy	**Autotrophic** organisms convert solar energy into chemical energy by photosynthesis. *Examples* Prokaryotae: blue-green algae Protoctista: seaweed Plants
Primary consumer	**Consumes** (eats) producers Digests and absorbs carbohydrates/lipids/proteins which contain energy	**Heterotrophic** organisms *Examples* Animals: sheep Protoctista: amoeba
Secondary consumer	**Consumes** (eats) primary consumers Digests and absorbs carbohydrates/lipids/proteins which contain energy	**Heterotrophic** organisms *Examples* Animals: dog
Decomposer	**Decomposes** dead organisms Digests and absorbs carbohydrates/lipids/proteins which contain energy	**Saprobiotic** organisms *Examples* Animals: earthworm Fungi: mushroom Prokaryotae: bacteria

ENERGY TRANSFER AND LOSSES BETWEEN TROPHIC LEVELS

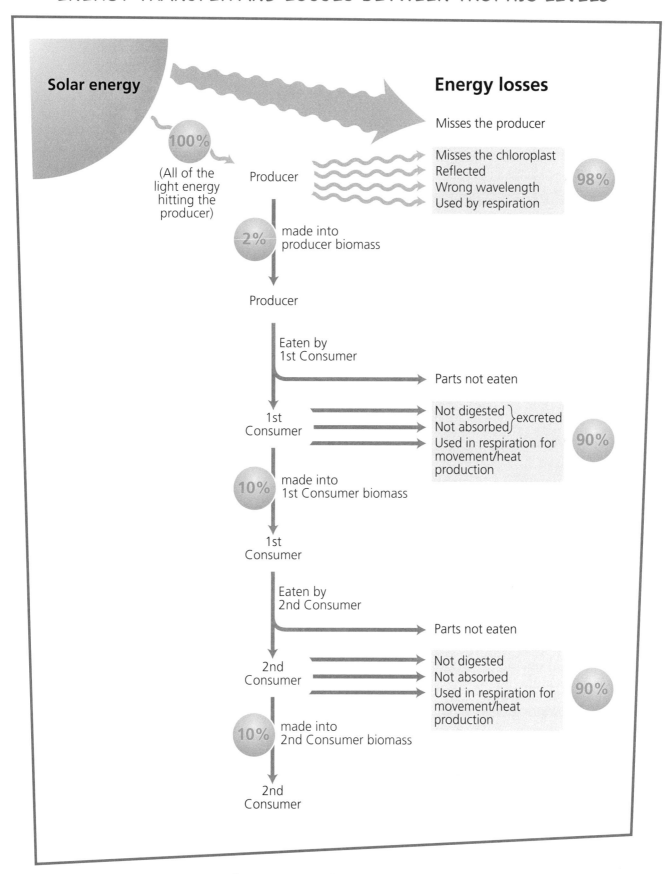

MUST REMEMBER

There are rarely more than four trophic levels because:
- only a limited amount of energy is available.
- energy is lost between each trophic level.

There is not enough energy left in the ecosystem to support more than **three** consumer levels.

THE CARBON CYCLE

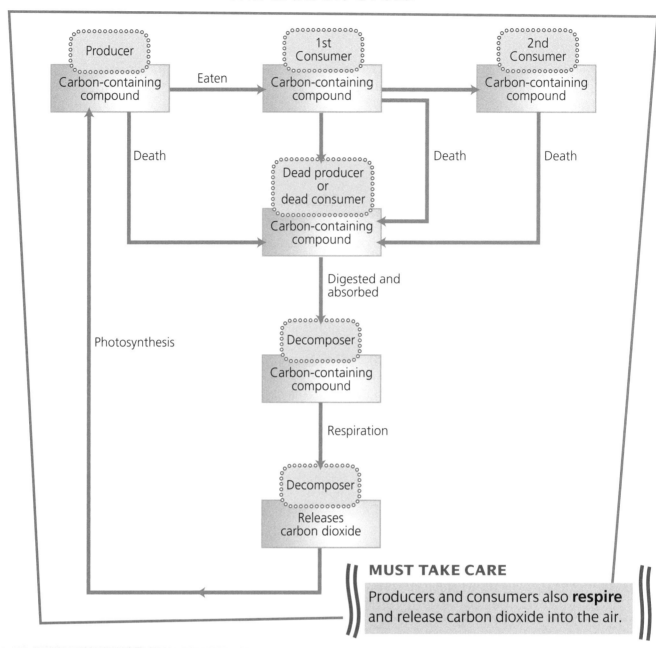

MUST TAKE CARE

Producers and consumers also **respire** and release carbon dioxide into the air.

Event in carbon transfer	Detail
Organic – **carbon-containing** – compounds are passed from one living thing to another.	• One organism eats another. • The organism digests and absorbs the organic molecule. • It is **assimilated** – included – into the body of the other organism.
An organism dies.	• **Decomposers** produce enzymes. • Decomposers digest the organic molecules and absorb them.
Carbon is passed into the environment as an inorganic molecule.	• Decomposers, producers and consumers respire. • Organic (carbon-containing) molecules are broken down – **hydrolysed**. • Energy (**ATP**) is released, water and **carbon dioxide** are produced.
Carbon is removed from the environment as an inorganic molecule.	• **Autotrophic** organisms (plants) absorb carbon dioxide. • During **photosynthesis**, carbon dioxide is converted into an organic molecule.

MUST TAKE CARE

Carbon passes from one living thing to another in the form of a **carbon-containing compound**, e.g. carbohydrates, lipids and proteins, and **not as carbon**.

NITROGEN CYCLE

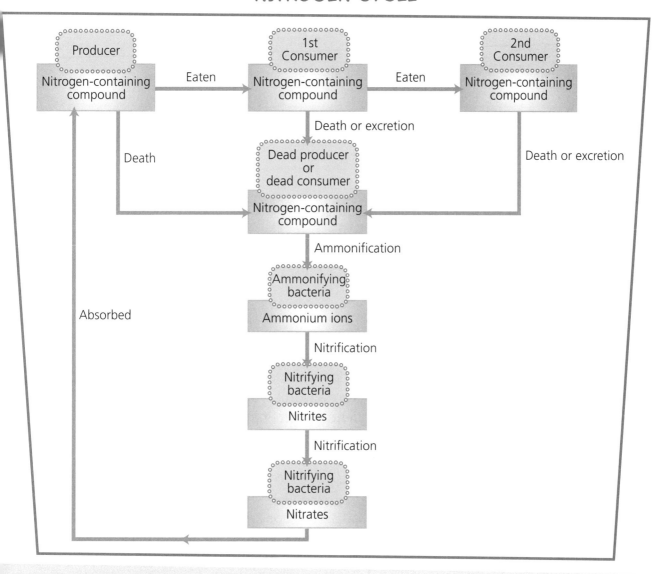

Event in nitrogen transfer	Detail
Organic **nitrogen-containing** compounds are passed from one living thing to another.	• One organism eats another. • Each organism digests and absorbs the organic molecule. • It is **assimilated** – included - into the body of the other organism.
After death, organic nitrogen-containing compounds are broken down into inorganic nitrogen-containing compounds by **bacteria**.	**Ammonification** • Proteins are broken down into ammonium ions. **Nitrification** • Ammonium ions are oxidised to nitrites. • Nitrites are oxidised to nitrates.
Inorganic nitrogen-containing compounds (nitrates) are absorbed by producers.	• A nitrogen atom from the nitrate joins with an organic molecule to make a protein.

OTHER ALTERNATIVES FOR NITROGEN

NITROGEN FIXATION
- In **nitrogen fixation**, nitrogen gas is 'fixed'.
- The enzyme **nitrogenase**, found in some bacteria, combines nitrogen and hydrogen to form ammonia.
- The nitrogen, 'fixed' as ammonia, is combined with organic acids to form bacterial protein.
- The bacterial protein joins the nitrogen cycle when the bacteria die.

DENITRIFICATION
- Some bacteria reduce nitrate ions to nitrogen gas.
- This removes usable nitrogen-containing compounds from the ecosystem.

SUMMARY OF BACTERIA IN THE NITROGEN CYCLE

Type of bacteria	Action
Ammonifying	They break down organic nitrogen-containing compounds to ammonium ions.
Nitrifying	They: • oxidise ammonium ions to nitrite ions. • oxidise nitrite ions to nitrate ions.
Nitrogen-fixing	They reduce nitrogen gas to ammonium ions.
Denitrifying	They reduce nitrate ions to gaseous nitrogen.

DEFORESTATION

Normal situation in forest	Effects
Nutrient cycling is continuous in a forest (see 'Carbon cycle', page 24, and 'Nitrogen cycle, page 25).	• Most nutrients are locked up in vegetation. • Dead material is broken down very fast – recycled. • Fast breakdown is due to high population of decomposers. • Nutrients are absorbed by trees. • Few nutrients remain in the soil.

Event	Effects
Trees are removed.	• The soil layer is thin. • It is no longer protected by trees. • The soil is washed or blown away. • Leaching removes any remaining nutrients. • This creates large barren areas.
Trees are removed and **crops** are planted.	• Crops absorb the limited nutrients in the soil. • Crops are removed (harvested). • Since plants are not left to decompose, nutrients are not available. • The yield falls over the years. • No artificial fertiliser is used – farmers are too poor. • Crops fail. (Land is abandoned.)

WHY CONSERVE RAINFORESTS?

A rainforest:
- is a **climax community**.
- is a stable ecosystem.
- has a high level of **biodiversity**.
- contains a huge range of **habitats** and **niches**.
- includes some unique **populations**.
- is a possible source of **drugs**.

In the processes that take place in a rainforest:
- the nutrient cycles involve many **producers**.
- **photosynthesis** removes carbon dioxide and adds oxygen to the air.
- there is a reduction in atmospheric carbon dioxide, so global warming is checked.

Removing trees in **deforestation** has these effects:
- Soil is exposed.
- **Leaching** takes place.
- **Erosion** occurs.
- So diversity falls.
- There are fewer producers, primary consumers and secondary consumers.
- This has a major effect on **food webs**.
- Some organisms may become **extinct**.

Having fewer producers causes:
- less carbon dioxide to be removed from the atmosphere.
- global warming to increase.

THE BIOCHEMISTRY OF PHOTOSYNTHESIS

The **basic equation** for **photosynthesis** is:

$$\text{carbon dioxide} + \text{water} \xrightarrow[\text{chlorophyll}]{\text{light energy}} \text{carbohydrate} + \text{oxygen}$$

This equation shows only the raw materials and **end products**, not what is happening.

LIMITING FACTORS

- Photosynthesis will only continue if the reaction conditions are favourable and if there is a sufficient supply of the raw materials.
- Any one of these that is not in abundance will reduce the **rate** of photosynthesis and is therefore called a **limiting factor**.

Limiting factor	Detail
Light intensity	• Light is the source of energy for photosynthesis.
Chlorophyll	• Chlorophyll is excited by light. • It converts solar energy into chemical energy.
Light wavelength	• Chlorophyll and other pigments are excited by particular wavelengths of light. • These are mainly red and blue.
Water	• Water is broken down by light. • It supplies the hydrogen to make carbohydrate. • Oxygen is released as a waste product.
Carbon dioxide	• Carbon dioxide is 'fixed' by combining with a 5-carbon sugar. • The process is controlled by enzymes.

THE TWO STAGES IN PHOTOSYNTHESIS

Stage	Diagram
Light-dependent stage Occurs in the **grana** of the chloroplast	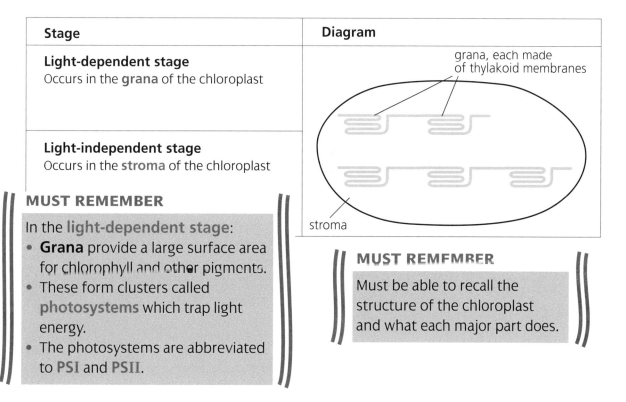 grana, each made of thylakoid membranes stroma
Light-independent stage Occurs in the **stroma** of the chloroplast	

MUST REMEMBER

In the **light-dependent stage**:
- **Grana** provide a large surface area for chlorophyll and other pigments.
- These form clusters called **photosystems** which trap light energy.
- The photosystems are abbreviated to **PSI** and **PSII**.

MUST REMEMBER

Must be able to recall the structure of the chloroplast and what each major part does.

LIGHT-DEPENDENT STAGE

The **light-dependent stage** of photosynthesis happens only in the presence of light.

Event and diagram	Detail
Light falls on chlorophyll 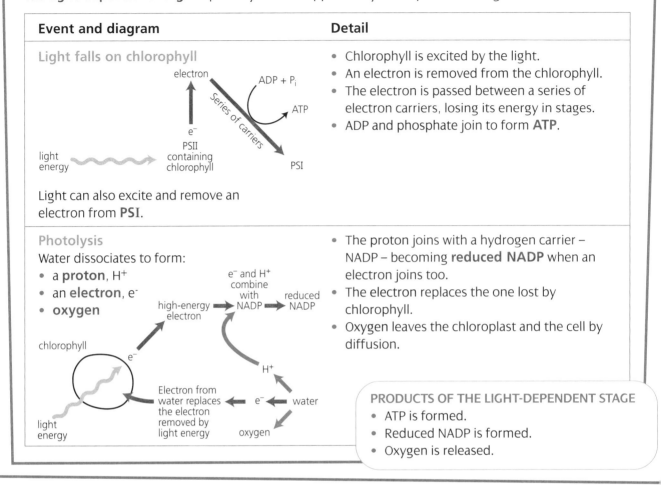 Light can also excite and remove an electron from **PSI**.	• Chlorophyll is excited by the light. • An electron is removed from the chlorophyll. • The electron is passed between a series of electron carriers, losing its energy in stages. • ADP and phosphate join to form **ATP**.
Photolysis Water dissociates to form: • a **proton**, H^+ • an **electron**, e^- • **oxygen**	• The proton joins with a hydrogen carrier – NADP – becoming **reduced NADP** when an electron joins too. • The electron replaces the one lost by chlorophyll. • Oxygen leaves the chloroplast and the cell by diffusion.

> **PRODUCTS OF THE LIGHT-DEPENDENT STAGE**
> • ATP is formed.
> • Reduced NADP is formed.
> • Oxygen is released.

LIGHT-INDEPENDENT STAGE

• The **light-independent stage** of photosynthesis happens with or without the presence of light, **but**
• it does require the products of the light-dependent stage.

Event and diagram	Detail
(1) Carbohydrate is formed 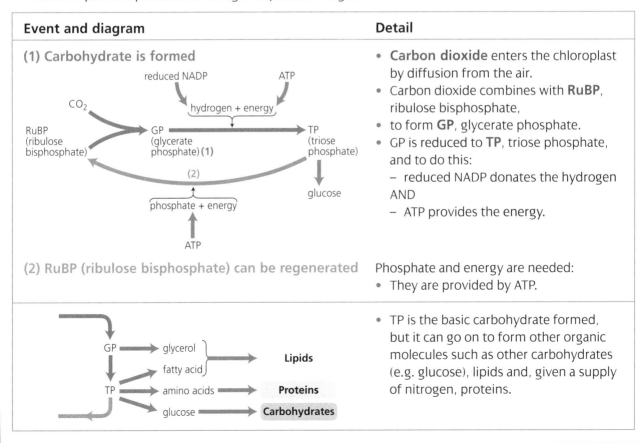	• **Carbon dioxide** enters the chloroplast by diffusion from the air. • Carbon dioxide combines with **RuBP**, ribulose bisphosphate, • to form **GP**, glycerate phosphate. • GP is reduced to **TP**, triose phosphate, and to do this: – reduced NADP donates the hydrogen AND – ATP provides the energy.
(2) RuBP (ribulose bisphosphate) can be regenerated	Phosphate and energy are needed: • They are provided by ATP.
	• TP is the basic carbohydrate formed, but it can go on to form other organic molecules such as other carbohydrates (e.g. glucose), lipids and, given a supply of nitrogen, proteins.

THE BIOCHEMISTRY OF RESPIRATION

Definition of **respiration**:

> Respiration involves breaking down large complex organic molecules, e.g. glucose, to produce a molecule that contains a small amount of energy – ATP.

ATP is an immediate source of energy:
- Only one step is necessary to release the energy.

$$\text{ATP} \rightleftharpoons \text{ADP} + \text{phosphate (P}_i\text{)} + \text{energy}$$

- The amount of energy released is small enough to be useful.

> **MUST TAKE CARE**
> - P represents the **element** phosphorus; it does NOT represent phosphate.
> - P_i represents inorganic phosphate.
>   Must either write out 'phosphate' in full or represent phosphate as P_i.

Carbohydrate is the main respiratory substrate.
- Other organic molecules, such as lipids and proteins, can be converted into forms that can also be used.

AEROBIC RESPIRATION

Stage and location	Diagram
(1) Glycolysis • occurs in the **cytoplasm** of the cell.	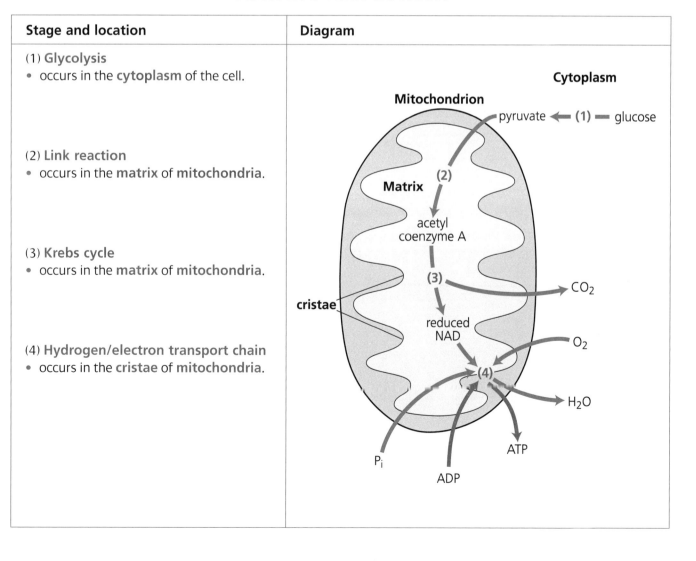
(2) Link reaction • occurs in the **matrix of mitochondria**.	
(3) Krebs cycle • occurs in the **matrix of mitochondria**.	
(4) Hydrogen/electron transport chain • occurs in the **cristae of mitochondria**.	

(1) GLYCOLYSIS

Chemical compound	Detail	ATP Loss	ATP Gain
Glucose (6 carbons) ↓ Glucose phosphate (6 carbons)	• ATP is broken down to ADP and P_i and energy. • This phosphate is used to **phosphorylate** glucose. • Phosphorylation activates glucose.	1	
↓ Fructose phosphate (6 carbons)	• Glucose is changed into its **isomer** fructose. • This changes the shape of the molecule so it can break more easily.		
↓ Fructose bisphosphate (6 carbons)	• ATP is broken down to ADP and P_i, and energy is released. • This phosphate is used to **phosphorylate** fructose phosphate into fructose bisphosphate.	1	
↓ Triose phosphate, TP (3 carbons)	• Fructose bisphosphate is **hydrolysed**, and • **Two 3-carbon** molecules are formed.		
	• The 3-carbon molecules are **oxidised**, with the removal of hydrogen. • The hydrogen is accepted by two molecules of NAD, forming **reduced NAD**. • 2ADP and $2P_i$ join, forming 2**ATP**.		2 × 2
↓ Pyruvate (3 carbons)		Totals 2	4
		Net gain: 2ATP	

MUST TAKE CARE

- Do **not** need to remember all the names of the stages in glycolysis in the table above.
- Information in the table is to reinforce some of the biochemistry and to identify what process is happening where.

For practice in answering A2 Biology questions, why not use *Collins Do Brilliantly A2 Biology*?

GLYCOLYSIS MADE SIMPLE

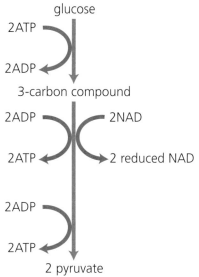

SUMMARY OF GLYCOLYSIS

Glucose is changed to pyruvate:
- 1 6-carbon molecule is changed to 2 3-carbon molecules.

Net ATP gain of two molecules:
- 2 ATP molecules are used.
- 4 ATP molecules are produced.

2 NAD molecules each receive a hydrogen to form 2 reduced NAD molecules.

- Pyruvate moves into the matrix of the mitochondria.

MUST REMEMBER

Must learn the level of detail in the Summary.

(2) LINK REACTION

Chemical compound	Detail
Pyruvate (**3 carbons**)	Hydrogen is released and accepted by **NAD** to form **reduced NAD**.
Acetyl coenzyme A (**2 carbons**)	A carbon atom is removed as carbon dioxide.

SUMMARY OF THE LINK REACTION

Since 2 pyruvate molecules are formed in glycolysis, therefore 2 go on to the link reaction.

Products formed are:
- 2 reduced NAD
- 2 carbon dioxide
- 2 acetyl coenzyme A

MUST TAKE CARE

- The formation of carbon dioxide has nothing to do with the oxygen that we breathe in.
- The carbon and the oxygen both come from the COOH group of the pyruvate molecule.
- So when CO_2 is lost, a hydrogen is always released, too.

(3) KREBS CYCLE

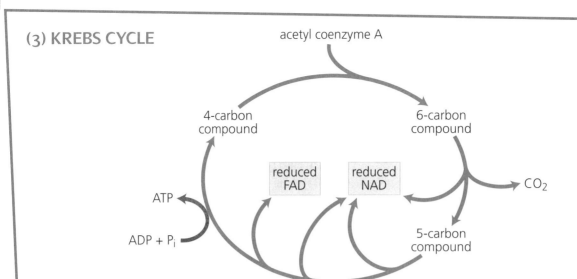

The reaction

Acetyl coenzyme A (**2 carbons**) combines with a **4-carbon** compound (oxaloacetate).

The **6-carbon** compound is broken down into a 5-carbon compound.

The **5-carbon** compound is broken down into the same **4-carbon** compound as before (oxaloacetate).

Products and by-products

The reaction forms a **6-carbon** compound (citrate).

The reaction releases:
- carbon in the form of carbon dioxide.
- hydrogen which combines with
  - **NAD** to form **reduced NAD**.

The reaction releases:
- carbon, in the form of carbon dioxide.
- hydrogen which combines with
  - **NAD** and
  - **FAD** (another coenzyme) to form **reduced NAD** and **reduced FAD**.
- There is enough energy released to join ADP and P_i to make **ATP**.

SUMMARY OF KREBS CYCLE

2 molecules of acetyl coenzyme A are formed and enter the Krebs cycle. This means that, starting with one molecule of glucose, the cycle happens twice. Products formed are:
- 6 reduced NAD
- 4 CO_2
- 2 reduced FAD
- 2 ATP

MUST REMEMBER

- Joining phosphate to another molecule is called **phosphorylation**.
- If the energy comes from the substrate, this is called **substrate level phosphorylation**, e.g. glycolysis and Krebs cycle.

WHAT HAS BEEN PRODUCED SO FAR?

Stage	ATP	Reduced NAD	Reduced FAD	CO_2
Glycolysis	net 2	2	0	0
Link	0	2	0	2
Krebs	2	6	2	4
Total	**4**	**10**	**2**	**6**

- Not much ATP has been formed directly – yet.
- Lots of hydrogen has been released and is in the form of either reduced NAD or reduced FAD.
- It does not matter where the reduced NAD or reduced FAD have been formed in aerobic respiration. These compounds will all give their hydrogen to the hydrogen/electron transport chain.

(4) HYDROGEN/ELECTRON TRANSPORT CHAIN

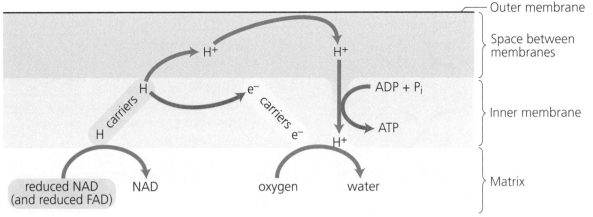

What happens	Detail
• **Reduced NAD** (and **reduced FAD**) delivers **hydrogen** to the inner membrane (**cristae**) of the mitochondrion. • **Hydrogen carriers** in the cristae combine with the hydrogen atoms.	Hydrogen atoms break down into: • protons (H^+). – protons pass into the space between the inner and outer membranes. • electrons (e^-). – electrons pass back to the inside of the inner membrane.
• Protons pass back into the matrix through a special channel in the membrane.	• An enzyme (cytochrome oxidase) attached to the membrane joins protons with electrons and oxygen to form water. • **Oxygen** is the **terminal hydrogen acceptor**.
• The special channel is associated with an enzyme, ATP synthetase.	• The flow of protons acts as a driving force. • ADP and phosphate are joined to make **ATP** by oxidative phosphorylation.

MUST REMEMBER

The process is called **oxidative phosphorylation** because
- phosphate is added to ADP, hence **phosphorylation**,
- the energy for the process is the result of hydrogen joining to oxygen, hence **oxidation**.

MUST TAKE CARE

Will not need to remember the numbers of reduced NAD or reduced FAD produced. These numbers just show where ATP molecules come from.

HOW MANY ATP MOLECULES ARE MADE?

Stage	ATP	Reduced NAD	Reduced FAD	CO_2
(1) Glycolysis	net 2	2	0	0
(2) Link	0	2	0	2
(3) Krebs	2	6	2	4
Total of each molecule type	4	10	2	6

- For every hydrogen that is given to the transport chain by reduced NAD, 3 molecules of ATP are formed.
- For every hydrogen that is given to the transport chain by reduced FAD, 2 molecules of ATP are formed.

| ATP formed | 4 | $10 \times 3 = 30$ | $2 \times 2 = 4$ | 0 |

These stages give a grand total of **38 molecules of ATP**.

MUST TAKE CARE

Oxidation occurs in respiration, but in:
- glycolysis
- link reaction
- Krebs cycle

oxidation is the **removal of hydrogen**.
Only in the hydrogen/electron transport chain is **oxygen needed**.

ANAEROBIC RESPIRATION

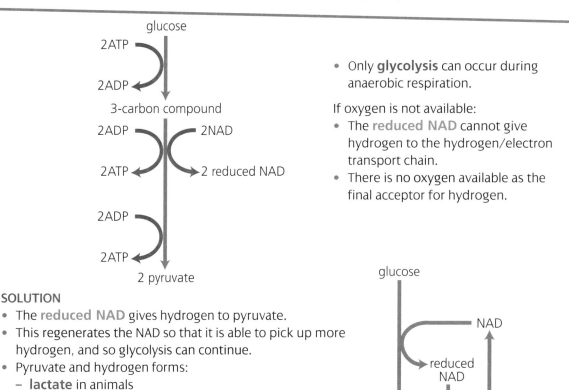

- Only **glycolysis** can occur during anaerobic respiration.

If oxygen is not available:
- The **reduced NAD** cannot give hydrogen to the hydrogen/electron transport chain.
- There is no oxygen available as the final acceptor for hydrogen.

SOLUTION
- The **reduced NAD** gives hydrogen to pyruvate.
- This regenerates the NAD so that it is able to pick up more hydrogen, and so glycolysis can continue.
- Pyruvate and hydrogen forms:
  - **lactate** in animals
  - **ethanol** and CO_2 in plants and yeast.
- Lactate is toxic and the body can only tolerate a limited amount.
- It must therefore be broken down.
- This can only be done when oxygen becomes available again.

AEROBIC AND ANAEROBIC RESPIRATION COMPARED

Feature	Aerobic respiration	Anaerobic respiration
Stages	Glycolysis Link reaction Krebs cycle Hydrogen/electron transport	Glycolysis
ATP produced from one molecule of glucose	38	2
Where it occurs	Cytoplasm and mitochondria	Cytoplasm
Differences between animals and plants	Identical	Animals – lactate is formed Plants – ethanol and carbon dioxide are formed

USING DIFFERENT RESPIRATORY SUBSTRATES

- The organic substance that is the starting point for respiration is called the **respiratory substrate**.
- One way to find which substrate is being used is to calculate the **respiratory quotient** or **RQ**.

$$RQ = \frac{\text{amount of } CO_2 \text{ produced}}{\text{amount of } O_2 \text{ consumed}}$$

RQ VALUES FOR SOME IMPORTANT SUBSTRATES
Lipids and fats   0.7
Proteins   0.9
Carbohydrates   1.0

If the RQ is greater than 1.0, anaerobic respiration must be involved.

HOMEOSTASIS

The definition of **homeostasis**:

> Homeostasis is the maintenance of a constant internal environment.

In all organisms, many **internal factors** need to be maintained. Here are just a few.

What factors are kept constant?	What would happen if they were NOT kept constant?
pH	• pH changes affect **enzyme activity**, so biochemical reaction rates would be reduced.
Temperature	• Proteins and other organic molecules are **denatured** by high temperatures.
Glucose	• Soluble molecules affect the **water potential** of body fluids. • Since glucose is needed as a respiratory substrate, a reduced concentration would impair **respiration**.
Water	• Water potential will be affected and cells could become **dehydrated**.

HOW CONSTANT IS CONSTANT?

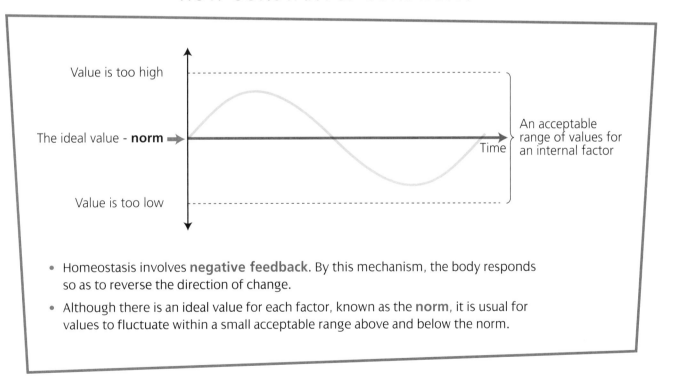

- Homeostasis involves **negative feedback**. By this mechanism, the body responds so as to reverse the direction of change.
- Although there is an ideal value for each factor, known as the **norm**, it is usual for values to fluctuate within a small acceptable range above and below the norm.

> **MUST REMEMBER**
>
> Negative feedback:
> - takes place if the value of an internal factor **rises or falls**.
> - returns the value towards the **norm**.

GENERAL PATTERN FOR ALL HOMEOSTATIC MECHANISMS

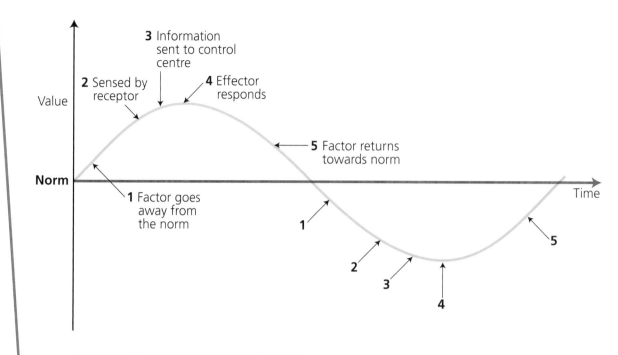

THE RESPONSE IS SENSITIVE AND RAPID

When values for a factor depart from the norm, the mechanism that responds to and controls a departure in one direction can be different from the response and control mechanism in the other direction.

- This makes the response more sensitive.
- It returns the factor towards the norm more rapidly.

For practice in answering A2 Biology questions, why not use *Collins Do Brilliantly A2 Biology*?

EXAMPLE 1: CONTROLLING BLOOD GLUCOSE LEVELS

Stage	When blood glucose level rises	When blood glucose level falls
1	Eating food containing glucose causes a **higher** blood glucose concentration (higher than the norm).	Using up glucose causes a **lower** blood glucose concentration (lower than the norm).
2	High glucose level is sensed by the **pancreas**.	Low glucose level is sensed by **pancreas**.
3	The hormone **insulin** is released by the **islets of Langerhans** within the pancreas: • Insulin travels in the **blood** to the **target cell**.	The hormone **glucagon** released by the **islets of Langerhans** within the pancreas: • Glucagon travels in the **blood** to the **target cell**.
4	**Liver** and **muscle** cells (target cells) respond to insulin: • The membrane of cells becomes **more permeable** to glucose. • Glucose moves **down** the concentration gradient, into the cell. • Cells **absorb** more glucose from the blood. • Also, an **enzyme** within those cells is activated. • The enzyme speeds up the **conversion** of **glucose** to **glycogen**. • This **lowers** the glucose concentration inside the cell. • Glycogen does not affect the water potential of the cell, whereas glucose does.	**Liver** cells (target cells) respond to glucagon: • An **enzyme** within those cells is activated. • This enzyme speeds up the conversion of **glycogen** to **glucose**. • This **raises** the glucose concentration inside the cell. • Glucose moves **down** the concentration gradient, out of the cell. • Therefore cells **release** glucose into the blood.
5	Blood glucose concentration **falls towards the norm**.	Blood glucose concentration **rises towards the norm**.

MUST TAKE CARE

Must be careful using the words beginning with the letter 'g':
- glucose
- glycogen
- glucagon

Must make sure to use them in the right place.

MUST REMEMBER

Glucose is a **soluble carbohydrate** found in the blood.
Glycogen is an **insoluble carbohydrate** found stored in the liver and muscle cells.
Glucagon is a **hormone** released by the islets of Langerhans.

EXAMPLE 2: CONTROLLING BODY TEMPERATURE

Stage	When body temperature rises	When body temperature falls
1	High air temperature or exercise causes the body temperature to **rise** (higher than the norm).	Low air temperature causes the body temperature to **fall** (lower than the norm).
2	This causes the **blood temperature** to **rise**. The higher temperature in blood is sensed by the **hypothalamus**.	This causes the **blood temperature** to **fall**. Lower temperature in blood is sensed by the **hypothalamus**.
3	**Nerve impulses** are sent from the hypothalamus to a number of **effectors** in: • blood vessels (arterioles) in skin, • sweat glands.	**Nerve impulses** are sent from the hypothalamus to a number of **effectors** in: • blood vessels (arterioles) in skin, • muscles attached to hairs in skin, • major organs.
4	The body responds to impulses by: • **vasodilation**, • increasing sweat.	The body responds to impulses by: • **vasoconstriction**, • hairs in skin standing on end, • increased metabolic activity.
5	Body temperature **falls towards the norm**.	Body temperature **rises towards the norm**.

MUST TAKE CARE

Vasodilation is the relaxation of:
- muscles in an **arteriole**,
- which carry blood to the **surface of the skin**.
- **More blood** travels to the surface.
- **More heat** is lost
- by **radiation**.

MUST REMEMBER

- Body temperature means blood temperature or **core temperature**.
- Many animals also have **behavioural mechanisms** to regulate temperature.

THE EYE

VERTICAL SECTION THROUGH THE EYE

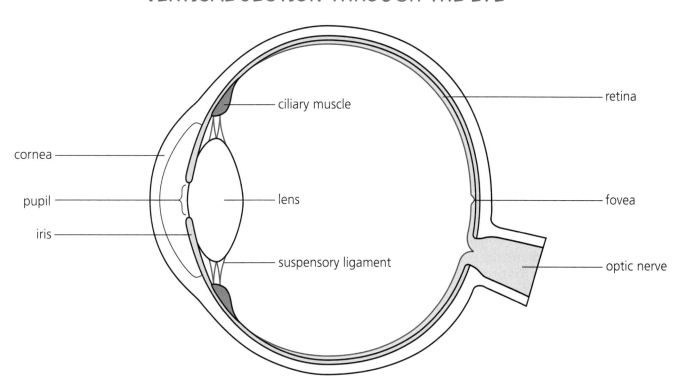

Part of the eye	What it is and its function
Cornea	• Clear section of the outer layer of the eye: – refracts light.
Iris	• Coloured part of the front of the eye: – controls the amount of light entering the eye.
Pupil	• Hole in iris: – allows light to pass through into the eye.
Lens	• Layered protein structure behind the pupil: – refracts light.
Retina	• Innermost layer: – is the light sensitive cell layer.
Fovea	• Small area on the retina: – where light is focused.

MUST REMEMBER

Must learn:
- the basic structure of the eye.
- the function of each part.

REFRACTION

As a **ray of light** moves from a less dense to a more dense medium, it bends. This process is **refraction**, and it takes place between the air and the cornea, and again at the lens.

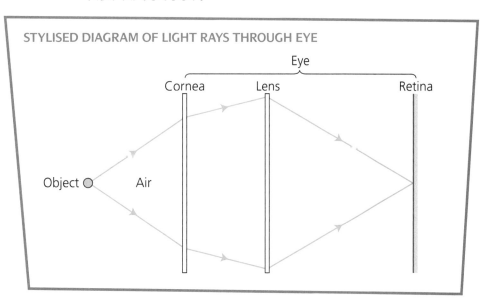

STYLISED DIAGRAM OF LIGHT RAYS THROUGH EYE

ACCOMMODATION

- Light from a point which enters the eye is **refracted** – bent – to hit a single point, the **fovea**, on the **retina**.
- This process is called **focusing**.
- The ability of the eye to focus both on near and then on distant objects is called **accommodation**.

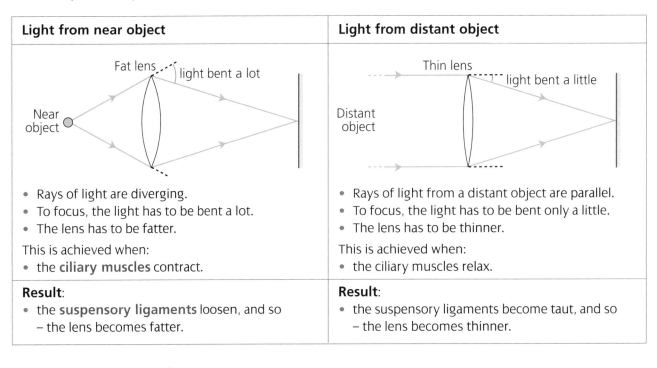

Light from near object	Light from distant object
- Rays of light are diverging. - To focus, the light has to be bent a lot. - The lens has to be fatter. This is achieved when: - the **ciliary muscles** contract.	- Rays of light from a distant object are parallel. - To focus, the light has to be bent only a little. - The lens has to be thinner. This is achieved when: - the ciliary muscles relax.

Result:
- the **suspensory ligaments** loosen, and so
  – the lens becomes fatter.
	Result:
	- the suspensory ligaments become taut, and so
	  – the lens becomes thinner.

MUST REMEMBER

Although it may seem wrong:
- When the ciliary muscles contract, the lens becomes fatter.

THE RETINA

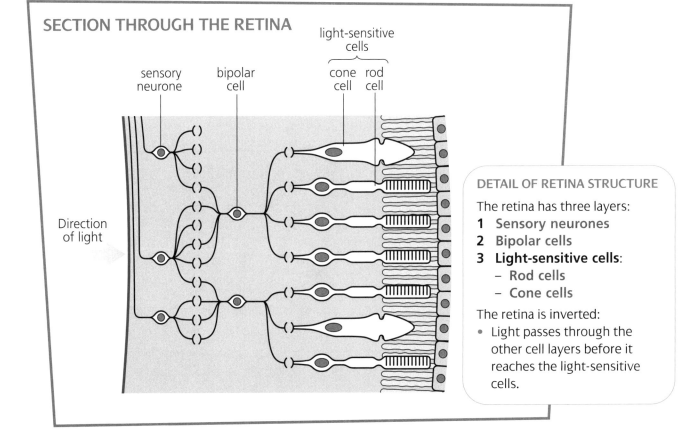

SECTION THROUGH THE RETINA

DETAIL OF RETINA STRUCTURE

The retina has three layers:
1. Sensory neurones
2. Bipolar cells
3. **Light-sensitive cells**:
   – Rod cells
   – Cone cells

The retina is inverted:
- Light passes through the other cell layers before it reaches the light-sensitive cells.

ROD CELLS

STRUCTURE OF A ROD CELL

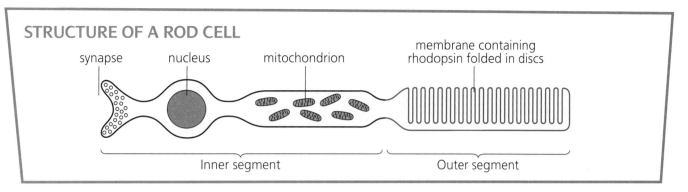

Property of rod cells	Detail
There is only one type of rod cell.	• The outer segment is densely packed with membrane-bound discs containing the light-sensitive pigment **rhodopsin**. • Rods are concentrated mostly in the edge regions of the retina. • There are no rods in the fovea.
They are extremely sensitive to light.	• Many rod cells synapse with a single bipolar cell: **convergence**. • The neurotransmitter released from many rods has an additive effect on the bipolar cell: **summation**.
Rods are responsible for vision in dim light.	• They do not support colour vision so, in very dim light, all objects appear in different shades of grey.

HOW LIGHT IS DETECTED

Rod cells contain the pigment **rhodopsin**, which is made of:
- **opsin** – a protein
- **retinal** – a light-sensitive molecule.

EVENTS WHEN LIGHT ENERGY REACHES THE RHODOPSIN

1. Retinal bends as it changes from one isomer to another.
    cis retinal → trans retinal
2. Rhodopsin breaks down as opsin and retinal break apart,
3. which causes a **generator potential**,
4. which causes an action potential to be produced,
5. which causes the rod to release a neurotransmitter into the synapse between the rod cell and bipolar cell.
6. An impulse is passed from the bipolar cell to a sensory neurone and to the brain.

CONE CELLS

Property of cone cells	Detail
There are three types in humans.	• Each type is sensitive to a different wavelength (colour) of light: **blue, red** or **green**.
Each type of cone contains a different pigment.	• Each pigment contains retinal and one of three different opsins.
The outer segment of cones is shorter than in rods.	• A cone contains less light-sensitive chemical than a rod, and so cones are less sensitive to light than rods.
Cone cells are concentrated mostly in the **fovea** of the retina.	• Cones are the **only type** of cell found in the **fovea**.
One bipolar cell may receive input from a single cone.	• This means that cone vision is more detailed: • We say that cones have a higher **visual acuity**.

Light stimulus → Impulse along four sensory neurones to brain — Sensory neurones — Bipolar cells — Synapses — Cones

The brain assesses the outputs of the three types of cones, and makes us aware of the resulting colour.

> **MUST TAKE CARE**
> - It is wrong to say 'red/blue/green cones'.
> - Must say they are cones 'which are sensitive to red/blue/green light'.

SUMMARY

Rods	Cones
• Black and white vision	• Colour vision
• Very sensitive to light	• High visual acuity
Rods enable us to see in dim light but only in black and white.	**Cones enable us to see clearly in colour but not in dim light.**

WATER TRANSPORT IN PLANTS

ROOT STRUCTURE

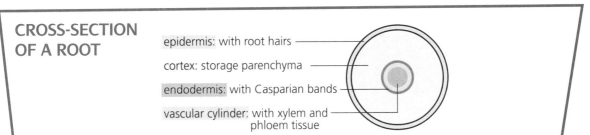

CROSS-SECTION OF A ROOT

- epidermis: with root hairs
- cortex: storage parenchyma
- endodermis: with Casparian bands
- vascular cylinder: with xylem and phloem tissue

Region of root	Function
Epidermis	Extensions of epidermal cells produce **root hairs**. Root hairs: • increase the surface area for absorption. • increase contact with soil water.
Cortex	It consists of irregularly shaped cells. • Spaces between cells are filled with water. Water can travel from epidermis to endodermis through intercellular spaces and **cell walls**: • The route is the **apoplastic pathway**. • Water travels down a water potential gradient. Water can also travel from cell to cell through the **cytoplasm**: • The route is the symplastic pathway. • Water passes between cells through **plasmodesmata**. • Water travels down a water potential gradient.
Endodermis	It is a single layer of cells. • Cell walls at right angles to the surface are thickened with a waterproof waxy strip, the **Casparian band**. • The Casparian band prevents water going through the endodermis by the apoplastic pathway. • Therefore water travels through the endodermis by the symplastic pathway.
Xylem	Xylem cells are elongated, hollow cells (like straws). • They have a secondary wall of **lignin**. • This stops cells collapsing when contents are under pressure. • Xylem carries water through the root into the stem.

SUMMARY OF WATER MOVEMENT INTO THE XYLEM

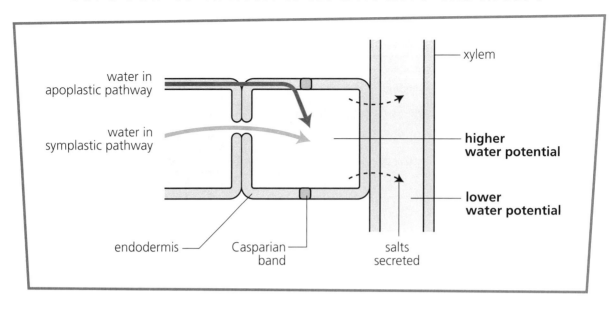

WATER MOVEMENT THROUGH THE XYLEM UP THE STEM

Mechanisms causing water movement	Explanation
Root pressure	Movement of water across the roots and into the xylem creates **root pressure**. • This mechanism will only move water a **short distance**. **Evidence** Water **exudes** from the top of a small plant when cut.
Cohesion-tension	Water molecules are **polar** – they 'stick' to one another. • This property is called **cohesion**. If one molecule moves, it pulls adjacent molecules. • This process creates **tension**. Water also 'sticks' to the walls of the xylem. • This property is called **adhesion**. • It **holds the column** of water together. **Evidence** The diameter of trees decreases during the day as water rises.
Transpiration	This is the **loss of water vapour** mainly through the stomata of leaves. • As water leaves the leaf it creates a **suction pressure**. • This **removes** water from the top of the xylem. **Evidence** Environmental factors that affect transpiration also affect water movement.

TRANSPIRATION

> In transpiration, water moves down a water potential gradient.

- The air inside the leaf is always saturated and therefore has a **higher** water potential than the air outside.
- Any environmental factor which produces a **lower** water potential will cause transpiration.

Factors affecting transpiration	Effect
Air movements	With **increased** air movement: • dry air is brought in contact with the leaf. • The air has a lower water potential than the air inside the leaf. • This creates a **greater** water potential gradient. **Result** More water can be removed.
Humidity	With **low** humidity: • air holds little water. • The air has a lower water potential than the air inside the leaf. • This creates a **greater** water potential gradient. **Result** More water can be removed.
Temperature	With **high** temperature: • water molecules gain kinetic energy. • They are more likely to move away from the plant. • This creates a **greater** water potential gradient. **Result** More water can be removed.
Light	With **high** light intensity: • stomata open. • More saturated air can escape the leaf. **Result** More water can be removed.

> For practice in answering A2 Biology questions, why not use *Collins Do Brilliantly A2 Biology*?

XEROPHYTES

A xerophyte is a plant adapted to survive in dry conditions.

Xerophytic adaptation	Effect
Thick waxy cuticle	Increased waterproofing
Needle-shaped leaves	Smaller surface area, **reduces** water loss
Stomata • reduced number • in pits • in grooves • only on bottom surface of leaf • shut during the day	**Reduces** transpiration
Leaves rolled	Smaller surface area, **reduces** water loss Stomata are inside the curled leaf, **reduces** water loss
Leaves reduced to spines	Smaller surface area, **reduces** water loss Trap layer of moist air, **reduces** water loss
Succulent stems	**Store** water
Roots widely spread	**Collect** any available water

MUST TAKE CARE

Cuticle only **reduces** water loss; it does **not** stop it.

MUST REMEMBER

Dry conditions are not only found in deserts.
- In cold areas, water will be frozen.
- Sandy areas will have little water since it drains away.
- Soil in sand dunes will have a lower water potential due to high salt levels, with the effect that water is removed from plants.

THE LIVER AND THE KIDNEY

BLOOD SUPPLY OF THE LIVER

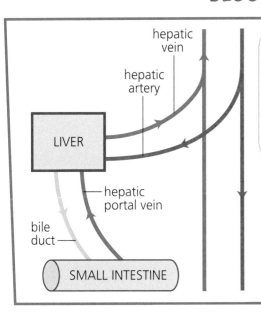

BLOOD VESSELS OF THE LIVER
- The **hepatic artery** brings oxygenated blood to the **liver**.
- The **hepatic vein** removes deoxygenated blood from the liver.
- The **hepatic portal vein** carries absorbed glucose and amino acids from the small intestine to the liver.
- The bile duct carries bile from the liver to the gall bladder, where it is then released into the small intestine to help in lipid digestion.

FUNCTIONS OF THE LIVER
Functions of the liver include:
- carbohydrate metabolism
- lipid metabolism
- protein metabolism
- bile synthesis

The first three functions are involved with **homeostasis**.

LIVER AS A HOMEOSTATIC ORGAN

SUBSTANCES WHOSE BLOOD LEVEL IS MAINTAINED BY THE LIVER

Substance	Detail
Carbohydrate/glucose • The **hepatic portal vein** contains an extremely variable glucose content. • Excess can be stored.	• **Glucose** in the liver is converted to **glycogen**. • Glycogen is an insoluble molecule that can be stored. • This conversion is stimulated by the hormone **insulin** (see 'Homeostasis', page 37): $$\text{Glucose} \xrightarrow{\text{insulin}} \text{Glycogen}$$ • **Glycogen** can be easily broken down to **glucose** again. • This conversion is to put glucose back in the blood. • It is stimulated by the hormone **glucagon**: $$\text{Glycogen} \xrightarrow{\text{glucagon}} \text{Glucose}$$
Lipid/fat • The liver is mainly involved with the processing of lipids rather than their storage.	• **Cholesterol** and **phospholipids** are removed from the blood and are broken down. • Excess **carbohydrates** are converted to fat.
Protein/amino acids • The **hepatic portal vein** contains an extremely variable amino acid content. • Excess cannot be stored.	**Deamination** • This is the process in which excess amino acids are broken down. • The **amino group** (NH_2) is removed. • This joins with a **hydrogen** (H), • to form **ammonia** (NH_3):  • The rest of the molecule is used for respiration. **Urea formation** • The ammonia produced in deamination is converted into a soluble excretory product – **urea**. • This occurs in a cyclic reaction called the **ornithine cycle**. $$CO_2 + NH_3 \xrightarrow{\text{Ornithine cycle}} \text{urea}$$

METHODS OF REMOVING NITROGENOUS WASTE

Nitrogenous waste is any waste product containing nitrogen.

Different organisms excrete different nitrogenous waste products:
- Uric acid – insects and birds
- Urea – mammals
- Ammonia – fish

What is released depends on its **toxicity** and water available to the organism.
- The more toxic, the more water is required.

NITROGENOUS WASTE REMOVAL IN DIFFERENT ORGANISMS

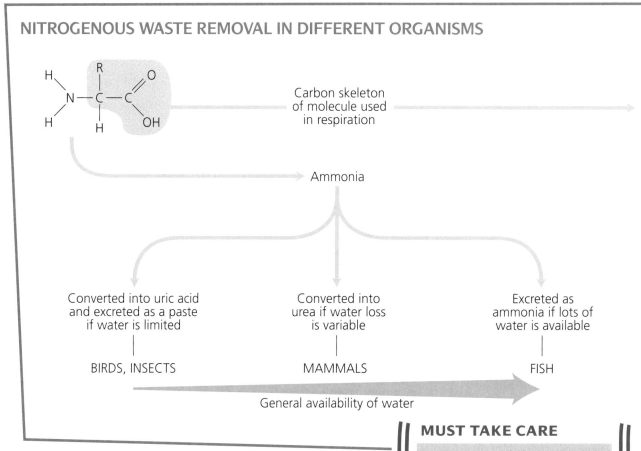

STRUCTURE OF THE KIDNEY

DIAGRAM OF A SECTION THROUGH A KIDNEY

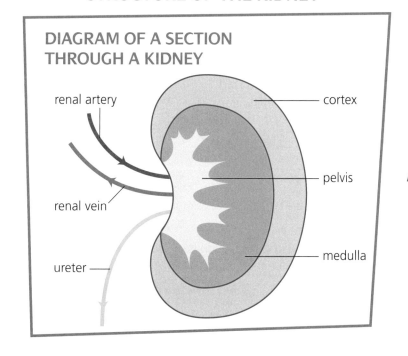

MUST TAKE CARE

Must not muddle up these:
- **Urea**, which is a nitrogenous (nitrogen-containing) waste product with the formula $(NH_2)_2CO$.
- **Urine**, which is the solution produced by the **kidney** containing urea and other solutes.

FUNCTIONS OF THE KIDNEY

- The **kidney** removes urea from the blood, which it receives in the renal artery.
  - The urea is removed in urine by the **ureter** that goes to the **bladder**.
- It also has a homeostatic function to balance the water content of the body.
- The working unit of the kidney is the **nephron**.

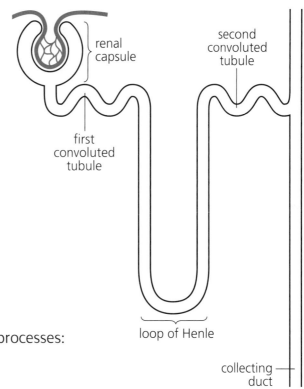

Kidney functions can be divided into 3 processes:
- **Ultrafiltration**
- **Reabsorption**
- **Concentration**

SUMMARY OF THE ROLES OF THE NEPHRON

Process	Where it happens	What happens
Ultrafiltration	Renal capsule	• **Filtration under pressure:** Small molecules are removed from the blood and pass into the **renal capsule**.
Reabsorption	First convoluted tubule	• **Molecules** that are small but useful are removed from the nephron and **returned to the blood.**
Concentration	Loop of Henle, second convoluted tubule and collecting duct	• Variable amounts of **water** are **returned to the blood**.

MUST REMEMBER

- Must be aware that parts of the nephron are named differently in different textbooks.
- Can use either of their names, noting that AQA specification A only uses the left terms in their questions:

AQA (A)	Alternatives
renal capsule	Bowman's capsule
first convoluted tubule	proximal convoluted tubule
second convoluted tubule	distal convoluted tubule

ULTRAFILTRATION

Ultrafiltration is filtering under pressure.

- The function of the kidney is to filter urea from the blood.
- This requires a filter with pores small enough to allow urea molecules through.

Structure	Function
The **afferent arteriole** taking blood to the **glomerulus** is wide.	To deliver lots of blood
The **efferent arteriole** taking blood from the glomerulus is narrow.	To create pressure in the glomerulus
Result: The **pressure** created forces small molecules into the nephron very rapidly.	
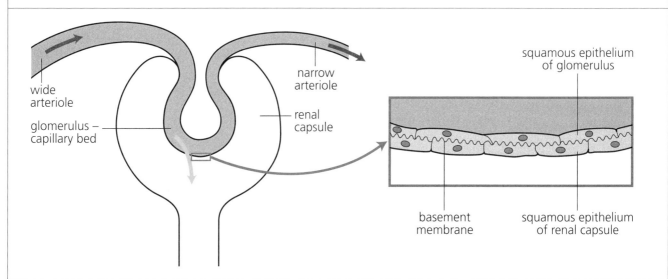	
The **squamous epithelium** of the glomerulus is thin.	To create a short diffusion pathway – no filter: allows both large and small molecules through
The **basement membrane** is the filter.	Only allows through molecules of a limited size
The squamous epithelium of the renal capsule is made of thin **podocytes** (cells with foot-like projections).	Allows the filtered molecules into the renal capsule with no barrier
Result: The **filter** created allows small molecules into the nephron, while large molecules stay in the blood.	

WHAT THE FILTRATE IN THE RENAL CAPSULE CONTAINS

Molecules the body must retain	Molecules the body must lose	Molecules the body must select
Glucose Amino acids	Urea	Water Mineral salts Hormones

Remaining in the blood after filtration are:
- blood cells, proteins and lipids.

MUST TAKE CARE

The liquid that goes through the nephron is called the **filtrate**. It is **not** urea or urine.

REABSORPTION

The **first convoluted tubule** is made of **cuboidal epithelial cells** that are modified to allow **reabsorption**.

The substances reabsorbed here are those that the body must retain. They include:

- glucose and amino acids.
  - In healthy animals, 100% of these are reabsorbed back into the blood.
  - This requires both **facilitated diffusion** and **active transport**.

Other substances are reabsorbed passively, down concentration gradients by **simple diffusion**. They include:

- water and salts.
  - Only 80% of these are reabsorbed.

Feature of epithelial cells	Function
Microvilli	• Increase surface area for diffusion
Mitochondria	• Provide ATP for active transport
Carrier proteins in membrane	• For facilitated diffusion and active transport

MUST REMEMBER

As molecules are removed from the nephron, capillaries remove them from the area round the epithelial cells.

CONCENTRATION

Concentration takes place in the loop of Henle, the second convoluted tubule and the collecting duct.
- At the start of the loop of Henle, the filtrate contains water, salts and urea.
- Most of the other substances have been reabsorbed.

Loop of Henle

The **descending limb** is permeable to water:
- Water leaves by osmosis.
- The water potential falls.

Result: The filtrate becomes more concentrated.

The **ascending limb** is impermeable to water:

First part:
- Salt (sodium and chloride ions) diffuses out,
- down a concentration gradient.

Second part:
- Chloride ions are actively transported out, followed by sodium ions,
- against a concentration gradient, because there is a lower concentration of salt inside the loop.

Result: The filtrate becomes less concentrated.

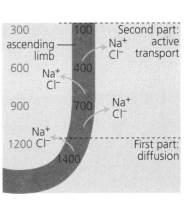

The movement of salts into the medulla of the kidney creates a **gradient** of salts within the **medulla**.

Second convoluted tubule	
Filtrate in the nephron has a lower salt concentration, thus a higher water potential, than the surrounding cortex cells.	Since the filtrate is **hypotonic** (less concentrated than the surrounding tissue): • water is reabsorbed by osmosis.
Collecting duct • It carries the filtrate through the medulla. • The medulla has an increasing concentration of salt (see diagram for loop of Henle above), thus a decreasing water potential.	• Water is reabsorbed from the collecting duct, • over its whole length.

Result: At the end of the nephron, concentrated urine (hypertonic) is formed – lots of urea in little water. This is now taken to the **bladder** in the **ureter**.

WATER HOMEOSTASIS

SUMMARY OF WATER INPUT AND OUTPUT

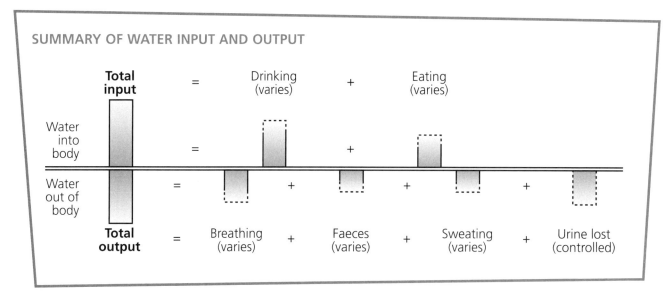

Water taken into the body must equal the water lost.
- The volume we drink and eat varies, as does the volume we lose by breathing, in faeces and by sweating.
- Only by varying the volume of water lost in urine can the water potential of the blood be controlled.

CONTROL OF WATER REABSORPTION

- The volume of water reabsorbed depends on the **water potential** of the blood.
- This is monitored by **osmoreceptors** in the **hypothalamus** of the brain.
- The hypothalamus stimulates the **pituitary gland** to secrete **ADH** (anti-diuretic hormone, ADH).
- ADH is produced all the time, but in varying amounts.

CONTROL OF BLOOD CONCENTRATION

Blood is too concentrated (has a low water potential)	Blood is too dilute (has a high water potential)
• More ADH is secreted into the blood.	• Less ADH is secreted into the blood.

- The target cells are:
  – the cells of the **second convoluted tubule**, **and**
  – the cells of the **collecting duct**.

MUST REMEMBER

ADH increases the permeability of the membrane of these cells.

• With more ADH, there is a high permeability.	• With less ADH, there is a low permeability.
• More water is reabsorbed.	• Less water is reabsorbed.
• Urine is very concentrated (hypertonic).	• Urine is very weak. (hypotonic)
• So less water is lost in urine.	• So more water is lost in urine.
Result **Water potential** of the blood rises.	**Result** Water potential of the blood falls.

DIGESTION

> **Definition:** Foods contain large organic insoluble molecules, such as proteins, carbohydrates and lipids. These are broken down by hydrolysis into small organic soluble molecules such as amino acids, sugars, fatty acids and glycerol.

Reason for digestion: So that the small organic soluble molecules can be **absorbed** and transported to every cell of the body.

STRUCTURE AND FUNCTION OF THE GUT WALL

The gut wall has three main layers:
- outer layer – **muscle**
- middle layer – **submucosa**
- inner layer – **mucosa**

Region of gut	Appearance	Special features and function
Oesophagus	longitudinal, circular, muscle, submucosa, mucosa	**Muscle** Two layers: longitudinal and circular – for **peristalsis**
Stomach	oblique, muscle, submucosa, mucosa, gastric pit	**Muscle** Three layers: longitudinal, circular and oblique – to mix food with enzymes **Mucosa** Thick layer with **gastric pits**. They contain 3 types of cell: • **Mucosal cells** secrete mucus. • **Oxyntic cells** secrete hydrochloric acid. • **Chief cells** secrete pepsinogen. **MUST TAKE CARE** Don't need to know the names of the cells but must know that the gastric pit is composed of three types of cell each producing something different.
Duodenum	muscle, submucosa, mucosa, villus	**Muscle** Two layers: longitudinal and circular – for **peristalsis** **Submucosa** Folded to form **villi** – increasing surface area for absorption • There are **enzymes** in the **membrane** of the epithelial cells. (See page 68.)
Ileum	muscle, submucosa, mucosa, villus	**Muscle** Two layers: circular and longitudinal – for **peristalsis** **Submucosa** **Villi** present Many **lymph** and **blood** vessels – absorb digested molecules

DIGESTIVE ENZYMES

STARCH DIGESTION

Part of gut	Secretion	Enzyme present	Substrate digested	Product	Notes
Buccal cavity	Saliva	**Amylase**	Starch	Maltose	Optimum pH is 7.
Stomach			No starch digestion here		
Duodenum	Pancreatic juice	**Amylase**	Starch	Maltose	Optimum pH is 8.
Ileum	See Notes, last column	**Maltase**	Maltose	Glucose	The enzyme is located in the plasma membrane of the epithelial cells which line the small intestine (duodenum and ileum).

PROTEIN DIGESTION

Part of gut	Secretion	Enzyme present	Substrate digested	Product	Notes
Buccal cavity			No protein digestion here		
Stomach	Gastric juice	**Pepsin** (**endopeptidase**)	Polypeptides	Peptides	• Pepsin is secreted in an inactive form – **pepsinogen**, • which is converted to its active form – **pepsin**, • by **hydrochloric acid**.
Duodenum	Pancreatic juice	**Trypsin** (**endopeptidase**)	Polypeptides	Peptides	• Trypsin is secreted in the inactive form – **trypsinogen**, • which is converted to its active form **trypsin**, • by the enzyme **enterokinase**.
Ileum	See Notes, last column	**Dipeptidase**	Dipeptide	Amino acids	The enzyme is located in the plasma membrane of the epithelial cells which line the small intestine (duodenum and ileum). The enzyme is also located in the epithelial cells of the small intestine.

LIPID DIGESTION

Part of gut	Secretion	Enzyme present	Substrate digested	Product	Notes
Buccal cavity			No lipid digestion here		
Stomach			No lipid digestion here		
Duodenum	Pancreatic juice	**Lipase**	Triglycerides (lipids)	Fatty acids and glycerol	Bile contains **bile salts** which emulsify the triglycerides.
Ileum			No lipid digestion here		

MUST REMEMBER

Bile is **not** an enzyme.
- Bile is produced by the liver and stored in the gall bladder.
- It contains:
  - an **alkali** (sodium hydrogencarbonate) which neutralises stomach acid,
  - **bile salts** which emulsify lipids.

ABSORPTION

When digestion is complete, the products – sugars, amino acids, fatty acids and glycerol – are absorbed from the small intestine into the blood or lymph system.

Digested product absorbed	How does it happen?
Amino acids and **glucose** move into a **blood vessel**.	**Amino acids and glucose:** • move from the small intestine by **facilitated diffusion**. • are linked to **sodium ions**. • move into the blood by **facilitated diffusion**. **Sodium ions:** • are removed by **active transport**.
Salts and **water** move into a **blood vessel**.	**Salts:** • move from the gut by **facilitated diffusion**. • move into the blood by **active transport**. **Water** moves down a water potential gradient created by salts: • moving from the gut by **osmosis**. • moving into the blood by **osmosis**.
Fatty acids and **glycerol** move into a **lymph vessel** – a **lacteal**.	**Fatty acids and glycerol:** • move from the gut by **diffusion**. They recombine in the epithelium cell to form a **triglyceride** (lipid). **Triglycerides:** • move into the lymph by **diffusion**.

MUST REMEMBER

As lipid is not soluble in water, it is coated by protein in the **lacteal** (lymph vessel in the centre of a villus), forming a chylomicron (**lipoprotein**). Lipid circulates in this form in the blood.

THE SMALL INTESTINE

- The **ileum** of the small intestine is the region of the gut in which most of the digested food is absorbed.
- The wall of the small intestine is specially adapted to form a very efficient **absorption surface**.

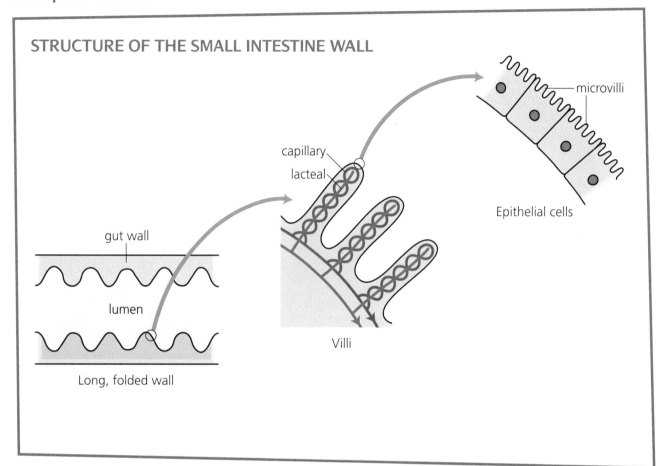

Feature of small intestine wall	Notes
The small intestine wall has a large surface area to allow **diffusion** and **active transport** to take place.	• The small intestine is long. • The wall is folded. • The folds are covered in **villi**. • The **epithelial cells** of the villi have membranes that are folded into **microvilli**. • The epithelial cell membranes contain many active transport **carrier molecules** that are involved in absorption.
The active transport system maintains a **large difference in concentration** of some substances either side of the epithelium.	• Muscles in the gut wall ensure that the contents of the small intestine are moving constantly. • The circulation of blood in the capillaries of the villi ensures that the absorbed amino acids and glucose are removed quickly from the intestine wall.
The intestine wall has a **thin exchange surface**. (Page 59 gives the themes for coloured type used above.)	• The epithelial cells give a short distance between the gut **lumen** and the blood.

RUMINANT DIGESTION

Ruminants are a group of mammals that eat plants – they are **herbivores** – and have a specialised digestive system.

Problem	Solution
Digesting carbohydrate • Some mammals (including cows) feed only on plants. • The commonest carbohydrate in their diet is **cellulose**. • Mammals do not produce **cellulase**, the enzyme that hydrolyses cellulose.	Cellulose — In plants eaten by cow; bacteria in gut produce cellulase ↓ Glucose — Used by bacteria ↓ Pyruvate — Produced by glycolysis/respiration in bacteria ↓ Fatty acid — Waste product; absorbed by cow • The association between ruminant and bacteria is a **mutualistic relationship**.
Obtaining enough protein • Grass as a food is low in protein. • It does contain other **nitrogen-containing compounds**.	Small proportion of protein + Non-protein nitrogen-containing compounds (e.g. ATP, DNA) — In plants eaten by cow Absorbed by bacteria ↓ Protein — Nitrogen-containing compounds made into protein by bacteria ↓ Protein — Digestion of protein + dead bacteria ↓ Amino acid — Absorbed by cow

> **MUST REMEMBER**
> In **mutualism**, two organisms are involved in a feeding relationship and both benefit.

CONTROLLING DIGESTION

- It is efficient to release enzymes only when required.
- Enzyme release is controlled by a combination of:
  - **nervous control**
  - **hormone control**.

WHY THE TWO SYSTEMS OF COORDINATION ARE NECESSARY

Nervous control is:	Hormone control is:
• quick • short lived • direct – affects only one area	• slower • long lasting • broadcast – affects more than one area
Any of the **reflex actions** below show these characteristics.	Any of the **hormones** below show these characteristics.

CONTROL IN DIFFERENT PARTS OF THE DIGESTIVE SYSTEM

Part of gut	Stimulus	Effect
Buccal cavity	• Smell or sight of food	• **Conditioned reflex** leads to secretion of saliva before food enters the buccal cavity.
	• Food in contact with the **taste buds** on the tongue	• **True reflex** leads to secretion of saliva onto the food.
Stomach	• Food in contact with the taste buds on the tongue	• **True reflex** leads to secretion of gastric juice.
	• Food in **stomach**	• The hormone **gastrin** is secreted. This causes the release of gastric juice into the stomach.

Pancreas and gall bladder	• Food in the **small intestine**	• The hormone **secretin** is secreted. This hormone causes the release of:
		  – alkaline fluid by the **pancreas**.
		• The hormone **CCK–PZ** is secreted. This hormone causes the release of:
		  – digestive enzymes by the pancreas,
		  – bile by the **gall bladder**, containing bile salts and alkaline fluid.

METAMORPHOSIS AND DIET

INSECT LIFE CYCLE

Stages of life cycle	Explanation
Adult lays **eggs**.	• So many eggs are laid to ensure survival that each egg has little stored food.
Eggs hatch into **larvae**.	• Eggs hatch after a short time, so the larvae are not fully developed. • Larvae need to eat to provide protein for growth and carbohydrates/sugars for energy.
Larva forms a **pupa**.	• During the pupal stage the body tissues are rearranged. • Reproductive organs develop and adult structures such as wings form.
Pupa hatches into an **adult**.	• The adults mate and the female moves around to lay eggs and thus distribute the offspring. • Adults need carbohydrates for energy.

Adult and larva are so different that the life cycle is said to show **metamorphosis**.

For practice in answering A2 Biology questions, why not use *Collins Do Brilliantly A2 Biology*?

METAMORPHOSIS

Since the diets are different in the different stages of the life cycle:

- the **mouthparts** of these stages are different,
- the **digestive enzymes** they produce are different.

Feature	Larval stage	Adult
	mandible	proboscis
Mouthparts	**Mandibles** are two blade-like structures that cut plant leaves.	**Proboscis** is a straw-like structure used to suck nectar.
Enzymes	• Protease • Lipase • Amylase • Sucrase • Maltase	• Sucrase

Changes in protein and energy requirements are associated with:

- **growth** in the larva,
- **reproduction** and **dispersal** in the adult.

ENZYMES, WHAT THEY DIGEST AND PRODUCTS

Enzyme	Substrate	Product	Use of products
Amylase	starch	maltose	All types of sugar for **energy**
Maltase	maltose	glucose	
Sucrase	sucrose	glucose + fructose	
Protease	protein	amino acids	Amino acids for **growth**
Lipase	lipids	fatty acids + glycerol	Fatty acids and glycerol for **growth** and **energy**

BEHAVIOUR

> Behaviour is the response to a stimulus.

A **stimulus** is a change in the environment:
- either internal (e.g. blood glucose level, temperature)
- or external (light, sound, temperature).

A **response** is the action taken.

THE SPINAL CORD AND SPINAL REFLEXES

- The simplest behaviour patterns are called **reflexes**.
- The nerve pathway is called a **reflex arc**.
- Reflex arcs that involve the spinal cord, but not the brain, are called **spinal reflexes**.

> **MUST REMEMBER**
> The **neurones** in a reflex arc:
> - **Sensory neurone** from the **sensory organ** to the **spinal cord**.
> - **Motor neurone** from the **spinal cord** to the **effector organ**.
> - **Relay neurone links** them together.

THE REFLEX ARC

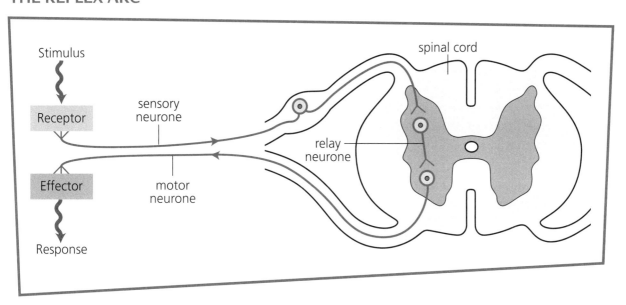

Description of a reflex response	Examples
• **Inborn** – inherited: not learnt • **Rapid** • **Repeatable**: always the **same response** to the **same stimulus** • Has **survival** value • Controlled by a nervous pathway called a **reflex arc**	blinking swallowing salivating

AUTONOMIC NERVOUS SYSTEM

The **autonomic nervous system** is part of the nervous system that controls internal glands and muscles that are beyond our conscious control. It has 2 divisions.

The **parasympathetic system** generally has an inhibitory effect:
- It prepares the body for rest and helps relaxation.
- The final neurotransmitter that the neurones release is acetylelcholine.

The **sympathetic system** generally has an excitatory effect:
- It prepares the body for activity and helps it to react to stress.
- The final neurotransmitter that the neurones release is noradrenaline.

EXAMPLES

Target organ	Effect of parasympathetic stimulation	Effect of sympathetic stimulation
Iris of eye	**Constricts** pupils	**Dilates** pupils
Intercostal muscles	**Decrease** breathing rate	**Increase** breathing rate
Heart	**Decrease** heart rate	**Increase** heart rate
Gut	**Stimulates** peristalsis	**Inhibits** peristalsis

SIMPLE BEHAVIOUR IN ANIMALS

The **behaviour** of some animals consists almost entirely of simple reflexes.
- The advantage is easy to see – it is **survival**.
- The behaviour cannot be adapted – it is **inborn**, i.e. **inherited**.

There are two types:
- **Taxis**
- **Kinesis**

TAXIS

A taxis is a directional response to a directional stimulus.

Examples
- Maggots move away from a source of light.
- Photosynthetic protoctista move towards light.

This behaviour allows animals to move towards condition which are favourable, and away from conditions which are not favourable.

KINESIS

A kinesis is a non-directional response to a non-directional stimulus.

Examples
- Woodlice cannot regulate their water loss. Therefore, in dry conditions they move faster and change their direction more frequently.

As a result of this behaviour, woodlice are more likely to move out of a dry area and find moist conditions.

THE KIDNEY

DEAMINATION

Deamination is the process by which excess amino acids are broken down.

In **deamination**:
- The **amino group** (NH_2) is removed.
- The amino group joins with a **hydrogen** (H),
- to form **ammonia** (NH_3).
- Ammonia is toxic and must be removed from the body.
- The rest of the molecule is used in respiration.

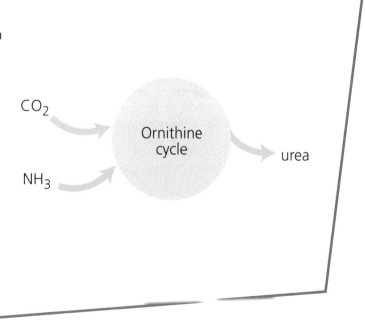

UREA FORMATION

- The ammonia produced in deamination is converted into a soluble excretory product – **urea**.
- Urea formation occurs in a cyclic reaction called the **ornithine cycle**.
- Urea is less toxic but it still must be removed.

STRUCTURE OF THE KIDNEY

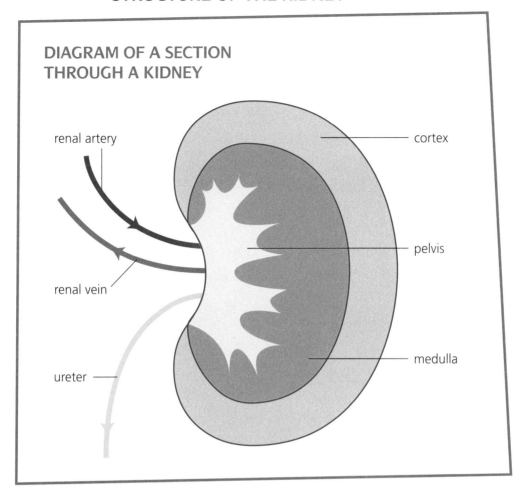

FUNCTIONS OF THE KIDNEY

- The **kidney** removes urea from the blood, which it receives in the renal artery.
  – The urea is removed in urine by the **ureter** that goes to the **bladder**.
- It also has a homeostatic function to balance the water content of the body.
- The working unit of the kidney is the **nephron** or **renal tubule**.

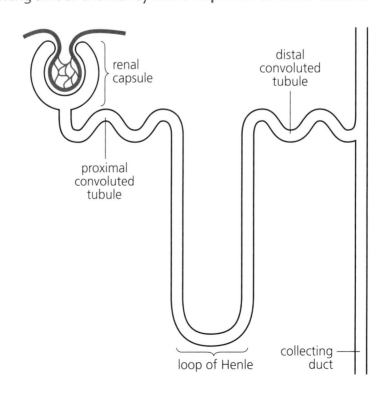

Kidney functions can be divided into 3 processes:
- **Ultrafiltration**
- **Reabsorption**
- **Concentration**

SUMMARY OF THE ROLES OF THE NEPHRON

Process	Where it happens	What happens
Ultrafiltration	Bowman's capsule	• **Filtration under pressure:** Small molecules removed from the blood and pass into the **renal capsule**.
Reabsorption	Proximal convoluted tubule	• **Molecules** that are small but useful are removed from the nephron and **returned to the blood**.
Concentration	Loop of Henle, distal convoluted tubule and collecting duct	• Variable amounts of **water** are **returned to the blood**.

MUST REMEMBER

- Must be aware that parts of the renal tubule are named differently in different textbooks.
- Must be able to use either of their names, noting that AQA specification B only uses the following terms in their questions:

AQA (B)	Alternatives
nephron	renal tubule
Bowman's capsule	renal capsule
proximal convoluted tubule	first convoluted tubule
distal convoluted tubule	second convoluted tubule

For practice in answering A2 Biology questions, why not use *Collins Do Brilliantly A2 Biology*?

ULTRAFILTRATION

Ultrafiltration is filtering under pressure.

- The function of the kidney is to filter urea from the blood.
- This requires a filter with pores small enough to allow urea molecules through.

Structure	Function
The **afferent arteriole** taking blood to the **glomerulus** is wide.	• To deliver lots of blood
The **efferent arteriole** taking blood from the glomerulus is narrow.	• To create pressure in the glomerulus
Result: The **pressure** created forces small molecules into the nephron very rapidly.	
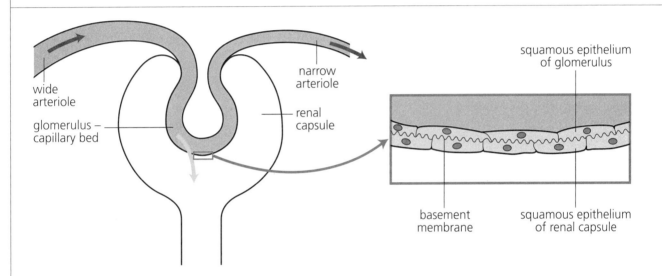	
The **squamous epithelium** of the glomerulus is thin.	• To create a short diffusion pathway – no filter: allows both large and small molecules through
The **basement membrane** is the filter.	• Only allows through molecules of a limited size
The squamous epithelium of the renal capsule is made of thin **podocytes** (cells with foot-like projections).	• Allows the filtered molecules into the renal capsule with no barrier
Result: The **filter** created allows small molecules into the nephron, while large molecules stay in the blood.	

WHAT THE FILTRATE IN THE RENAL CAPSULE CONTAINS

Molecules the body must retain	Molecules the body must lose	Molecules the body must select
Glucose Amino acids	Urea	Water Mineral salts Hormones

Remaining in the blood after filtration are:
- blood cells, proteins and lipids.

MUST TAKE CARE

The liquid that goes through the nephron is called the **filtrate**. It is **not** urea or urine.

REABSORPTION

The **proximal convoluted tubule** is made of **cuboidal epithelial cells** that are modified to allow for **reabsorption**.

The substances reabsorbed here are those that the body must retain. They include:
- glucose and amino acids.
  - In healthy animals, 100% of these are reabsorbed back into the blood.
  - This requires both **facilitated diffusion** and **active transport**.

Other substances are reabsorbed passively, down concentration gradients by **simple diffusion**. They include:
- water and salts.
  - Only 80% of these are reabsorbed.

Feature of epithelial cells	Function
Microvilli	• Increase surface area for diffusion
Mitochondria	• Provide ATP for active transport
Carrier proteins in membrane	• For facilitated diffusion and active transport

MUST REMEMBER
As molecules are removed from the nephron, capillaries remove them from the area round the epithelial cells.

CONCENTRATION

Concentration takes place in the loop of Henle, the distal convoluted tubule and the collecting duct.
- At the start of the loop of Henle, the filtrate contains water, salts and urea.
- Most of the other substances have been reabsorbed.

Loop of Henle

The **descending limb** is permeable to water:
- Water leaves by osmosis.
- The water potential falls.

Result: The filtrate becomes more concentrated.

The **ascending limb** is impermeable to water:

First part:
- Salt (sodium and chloride ions) diffuses out,
- down a concentration gradient.

Second part:
- Chloride ions are actively transported out, followed by sodium ions,
- against a concentration gradient, because there is a lower concentration of salt inside the loop.

Result: The filtrate becomes less concentrated.

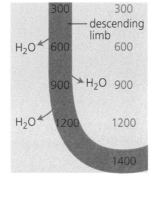
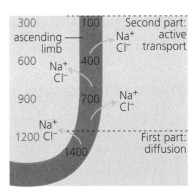

The movement of salts into the medulla of the kidney creates a **gradient** of salts within the **medulla**.

Distal convoluted tubule
Filtrate in the nephron has a lower salt concentration, thus a higher water potential, than the surrounding cortex cells.

Since the filtrate is **hypotonic** (less concentrated than the surrounding tissue);
- water is reabsorbed by osmosis.

Collecting duct
- It carries the filtrate through the medulla.
- The medulla has an increasing concentration of salt (see diagram for loop of Henle above), thus a decreasing water potential.

- Water is reabsorbed from the collecting duct,
- over its whole length.

Result: At the end of the nephron, concentrated urine (hypertonic) is formed – lots of urea in little water. This is now taken to the **bladder** in the **ureter**.

CONTROL OF WATER REABSORPTION

- The volume of water reabsorbed depends on the **water potential** of the blood.
- This is monitored by **osmoreceptors** in the **hypothalamus** of the brain.
- The hypothalamus stimulates the **pituitary gland** to secrete **ADH** (anti-diuretic hormone).
- ADH is produced all the time but in varying amounts.

CONTROL OF BLOOD CONCENTRATION

Blood is too concentrated (has a low water potential)	**Blood is too dilute** (has a high water potential)
• More ADH is secreted into the blood.	• Less ADH is secreted into the blood.

• The target cells are:
  – the cells of the **distal convoluted tubule**, and
  – the cells of the **collecting duct**.	

Result ADH increases the permeability of the membrane of these cells.	
• With more ADH, there is a high permeability.	• With less ADH, there is a low permeability.
• More water is reabsorbed.	• Less water is reabsorbed.
• Urine is very concentrated (hypertonic). • So less water is lost in urine.	• Urine is very weak (hypotonic). • So more water is lost in urine.
Result **Water potential** of the blood rises.	**Result** Water potential of the blood falls.

MUST TAKE CARE

Must not confuse these terms which appear similar:
- Urea – a molecule produced by the liver which is removed from the blood by the kidney.
- Urine – a liquid produced by the kidney containing urea.

MUST REMEMBER

Must not write simply that water is reabsorbed. To get marks, must use the correct terms, i.e. Water moves from a high**er** water potential to a low**er** water potential by **osmosis**.

THE BRAIN

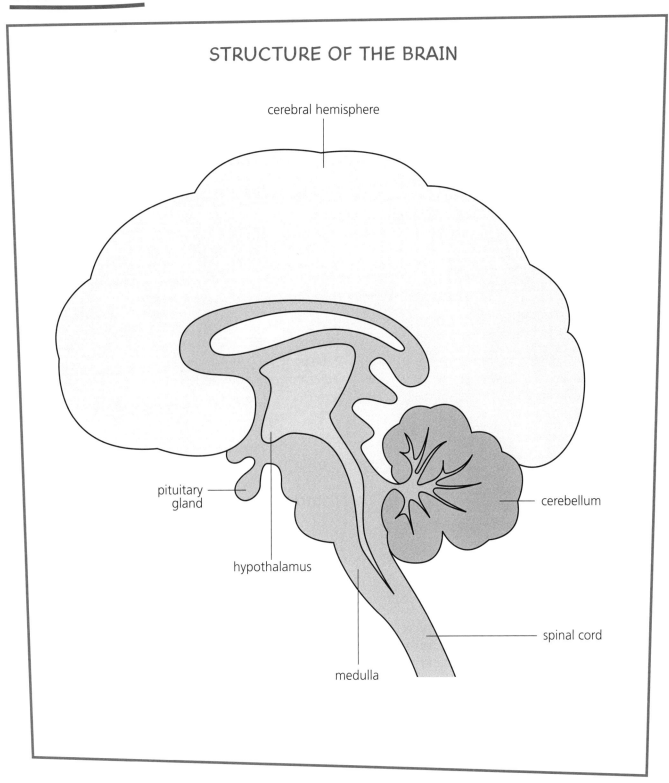

The **brain** is divided into three areas.

Hindbrain
- **cerebellum**, coordinating muscle contraction, and
- **medulla**, coordinating heartbeat and breathing.

Midbrain
- **hypothalamus**, linked to
- the **pituitary gland**, coordinating nervous and hormonal control.

Forebrain
- **cerebrum** divided into two **cerebral hemispheres**, right and left, connected by nerve fibres of the **corpus callosum**.

CEREBRAL HEMISPHERES

The surface of the cerebrum is folded and covered with a thin layer of neurone cell bodies, known as **grey matter** or **cerebral cortex**.

AREAS OF THE CEREBRAL CORTEX

Sensory areas	Association areas	Motor areas
• **Receive input** by sensory neurones from receptors in skin, muscles, eyes, ears and nose	• **Receive impulses** from the sensory areas • **Interpret the information**, relating it to previous experience • **Send out impulses** through the motor area and **initiate a response** • **Link** the sensory and motor areas of the brain	• **Send impulses** to muscles and glands
• The size of each area is related to the number of receptors it deals with.	• The areas are sites of memory, learning, reasoning and therefore intelligence.	• Each motor area is linked to one part of the body.

- One side of the brain controls the **opposite side** of the body.
- In the medulla the nerve cell fibres cross over so that information from the left cerebral cortex innervates the right side of the body, and vice versa.

VISUAL ASSOCIATION AREA

Information from the eyes is converted into 'vision'.

MUST REMEMBER
Although visual information is received by eyes, we 'see' with our brain!

Input from eyes	Brain associates input with...	After association...
Examples: • light intensity • colour • movement • shape	other sources such as: • input from other receptors. • past experiences/memory.	• impulses are sent to motor areas which initiate a response.

SPEECH AREAS

Locations of the speech areas	Roles of the speech areas
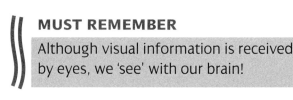	**Auditory sensory area** • receives input from ears.
	Auditory association area • interprets sounds into broad categories.
	Speech association area • makes sense of words.
	Speech motor centre • initiates speech.

REPRODUCTION

MALE AND FEMALE REPRODUCTIVE SYSTEMS

The male and female reproductive systems are very different. Each has several functions.

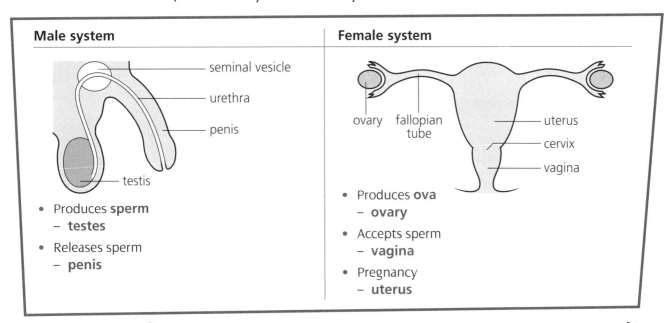

Male system
- Produces **sperm**
  - **testes**
- Releases sperm
  - **penis**

Female system
- Produces **ova**
  - **ovary**
- Accepts sperm
  - **vagina**
- Pregnancy
  - **uterus**

MUST TAKE CARE
Must be aware that the many structures in the female and male reproductive system will not be tested in the exam, but must know the names and functions of the parts.

PROCESSES IN THE REPRODUCTIVE ORGANS

Producing **gametes** (ova and sperm) is a major function.
- The process is **gametogenesis**.
- Gametogenesis is different in male and female, but the main stages are the same.

SPERMATOGENESIS – MAKING SPERM

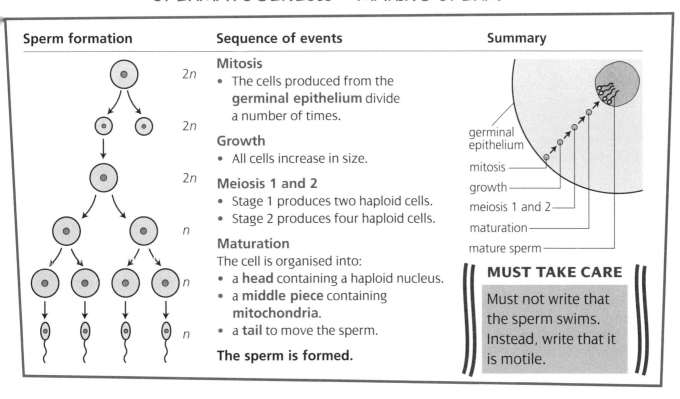

Sperm formation (diagram shows 2n cell dividing through stages to produce n sperm cells)

Sequence of events

Mitosis
- The cells produced from the **germinal epithelium** divide a number of times.

Growth
- All cells increase in size.

Meiosis 1 and 2
- Stage 1 produces two haploid cells.
- Stage 2 produces four haploid cells.

Maturation
The cell is organised into:
- a **head** containing a haploid nucleus.
- a **middle piece** containing **mitochondria**.
- a **tail** to move the sperm.

The sperm is formed.

Summary: germinal epithelium, mitosis, growth, meiosis 1 and 2, maturation, mature sperm.

MUST TAKE CARE
Must not write that the sperm swims. Instead, write that it is motile.

SPERM STRUCTURE

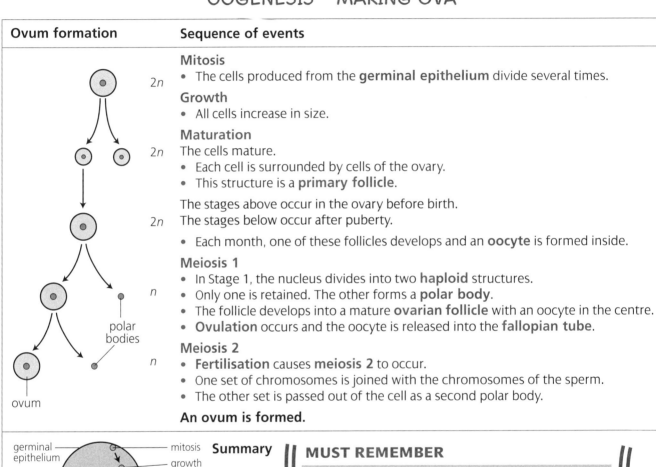

MOVEMENT OF SPERM

Several factors help move the sperm toward the oocyte:

1. **Ejaculation** pushes the sperm to the top of the **vagina**.
2. Sperm are motile and use their tail to move through the female tract.
3. The **uterus** contracts, pulling the sperm upwards.

Time scale
- Sperm can reach the fallopian tube in 4 to 6 hours.
- Sperm can remain alive in the female reproductive system for 48 hours.

OOGENESIS – MAKING OVA

Ovum formation	Sequence of events
	Mitosis • The cells produced from the **germinal epithelium** divide several times. **Growth** • All cells increase in size. **Maturation** The cells mature. • Each cell is surrounded by cells of the ovary. • This structure is a **primary follicle**. The stages above occur in the ovary before birth. The stages below occur after puberty. • Each month, one of these follicles develops and an **oocyte** is formed inside. **Meiosis 1** • In Stage 1, the nucleus divides into two **haploid** structures. • Only one is retained. The other forms a **polar body**. • The follicle develops into a mature **ovarian follicle** with an oocyte in the centre. • **Ovulation** occurs and the oocyte is released into the **fallopian tube**. **Meiosis 2** • **Fertilisation** causes **meiosis 2** to occur. • One set of chromosomes is joined with the chromosomes of the sperm. • The other set is passed out of the cell as a second polar body. **An ovum is formed.**

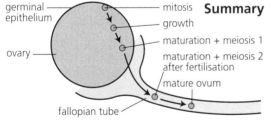

Summary
- mitosis
- growth
- maturation + meiosis 1
- maturation + meiosis 2 after fertilisation
- mature ovum

MUST REMEMBER

The stages of the cells during gametogenesis are all given different names to identify them as different. Need not remember the names; just the principles of the process.

WHAT HAPPENS IN THE OVARY

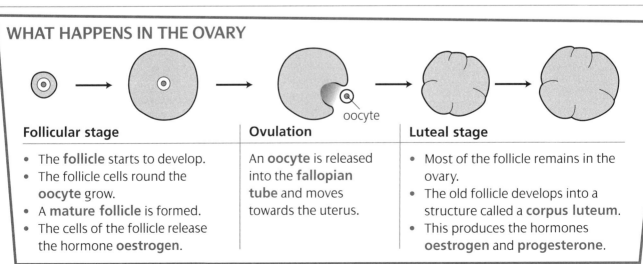

Follicular stage
- The **follicle** starts to develop.
- The follicle cells round the **oocyte** grow.
- A **mature follicle** is formed.
- The cells of the follicle release the hormone **oestrogen**.

Ovulation

An **oocyte** is released into the **fallopian tube** and moves towards the uterus.

Luteal stage
- Most of the follicle remains in the ovary.
- The old follicle develops into a structure called a **corpus luteum**.
- This produces the hormones **oestrogen** and **progesterone**.

FERTILISATION

- When the **oocyte** is released from the ovary it can then be fertilised by a **sperm** in the fallopian tube.

> Fertilisation is the fusion of sperm and oocyte.

Event	More detail	Diagram
Meeting of sperm and oocyte	• The oocyte is released from the ovary into the fallopian tube. • The sperm move through the vagina, uterus and into the fallopian tube.	oocyte, many sperm, jelly surrounding oocyte
Acrosome reaction	• **Calcium ions** enter the head of the sperm. • The **acrosome membrane** fuses with the sperm cell membrane. • **Hydrolytic enzymes** are released from the acrosome. • Enzymes digest the jelly surrounding the oocyte.	membrane of oocyte, jelly
Fusion of genetic material	• Membranes of the sperm head and the oocyte fuse. • The nucleus of the sperm enters the oocyte. • A **zygote** is formed.	membrane of oocyte, membrane of sperm head, sperm nucleus enters oocyte

FROM FERTILISATION TO IMPLANTATION

Event	More detail	Diagram
Cleavage	• Nuclear membranes of both haploid nuclei break down. • Chromosomes replicate. • Mitosis and cytokinesis takes place. • Many new cells are made.	
Blastocyst formation	• A ball of cells is made. • Outer cells tuck in and a hollow ball, the **blastocyst**, is formed. • Nutrition is provided by secretions of the uterus.	trophoblast: outer layer of blastocyst
Implantation	• The **trophoblast**, the outer layer of the blastocyst, grows into the **endometrium** of the uterus. • It forms **trophoblastic villi** which digest the endometrium.	blastocyst, endometrium of uterus, trophoblastic villus
Placenta formation	• The mother's blood vessels surround the villi. • Blood spaces are formed. • Each villus contains many capillaries.	maternal blood space, epithelium of villus, fetal capillary, epithelium of capillary, connective tissue

> **MUST REMEMBER**
>
> The blood of the fetus and mother are kept separate because:
> - Blood pressure is higher in the mother's body.
> - The blood group of the fetus may be different from the mother's.

THE PLACENTA

PLACENTAL FUNCTION

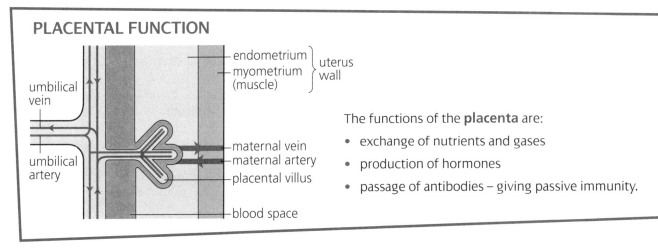

The functions of the **placenta** are:
- exchange of nutrients and gases
- production of hormones
- passage of antibodies – giving passive immunity.

HOW MATERIALS ARE PASSED ACROSS THE PLACENTA

Diffusion	Facilitated diffusion	Active transport	Exocytosis
Oxygen Carbon dioxide Urea	Glucose	Amino acids	Antibodies

HOW THE PLACENTA IS ADAPTED FOR EFFICIENT EXCHANGE

Feature	Adaptation
Large surface area	• The placenta is folded into **villi**. • The villi have **microvilli** on their membranes.
Maintenance of a concentration gradient	• A continuous supply of oxygenated blood is provided by the mother. • Oxygenated blood is removed from the placenta by the fetus through the **umbilical vein**. • The blood of the mother flows in the opposite direction to that of the fetus. • **Fetal haemoglobin** has a much higher affinity for oxygen than adult haemoglobin.
Short diffusion pathway	• Because maternal blood vessels are broken down, there is a reduction in the number of cell layers between the mother's blood and the fetal blood.

HORMONES PRODUCED BY THE PLACENTA

Hormone	Function
Human chorionic gonadotrophin, hCG	• It maintains the corpus luteum which will continue to produce oestrogen and progesterone.
Oestrogen	• It inhibits the production of FSH, so no new follicles develop.
Progesterone	• It maintains the endometrium (the lining of the uterus).

MUST REMEMBER

- The hormone hCG is only produced while the placenta is developing.
- The **corpus luteum** therefore degenerates after 3 months.
- The placenta takes over the production of oestrogen and progesterone.

HORMONE CONTROL OF BIRTH

Towards the end of pregnancy:
- the level of **progesterone** in the blood falls.
- the level of **oestrogen** in the blood **rises**.
  - This makes the uterus more sensitive to the hormone **oxytocin**.
  - Oxytocin causes the muscles in the uterus to contract.
  - These contractions will push the baby out of the uterus.

HORMONE CONTROL OF LACTATION

During pregnancy, oestrogen and progesterone stimulate the development of milk-producing tissue in the breasts.
↓
At birth, due to low levels of progesterone, the hormone **prolactin** from the pituitary is no longer inhibited.
↓
Prolactin stimulates the breasts to produce milk.
↓
When the baby suckles, this stimulates the pituitary gland to produce **prolactin** and **oxytocin**.
↓
Oxytocin stimulates the **muscles** in the walls of the milk ducts to contract, squeezing milk into the baby's mouth.
↓
Prolactin stimulates the production of more milk.

SUMMARY OF THE HORMONES INVOLVED IN MILK PRODUCTION

Hormone	Source and function
Oestrogen	• It is produced by the corpus luteum and placenta. • It stimulates the development of milk-producing tissue in breasts.

Progesterone	• It is produced by the corpus luteum and placenta.
	• It also stimulates the development of milk-producing tissue in breasts.
	  – A high concentration of progesterone inhibits prolactin.
	  – A low concentration of progesterone releases this inhibition.

Prolactin	• It is released from the pituitary gland when the breast is suckled. • It stimulates the breast to produce milk.
Oxytocin	• It is released from the pituitary gland when the breast is suckled. • It stimulates the muscles of the breast to contract, releasing milk.

CHANGES IN BLOOD CIRCULATION AROUND THE HEART

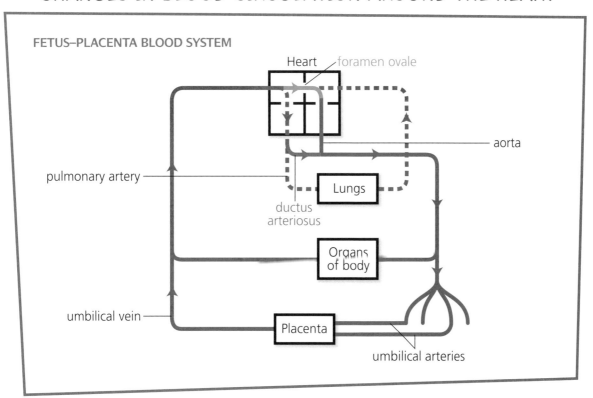

FETUS–PLACENTA BLOOD SYSTEM

CHANGES IN BLOOD CIRCULATION AROUND THE HEART

In adult	In fetus
• Blood is **deoxygenated** as organs of the body remove oxygen for aerobic respiration.	• Blood is **oxygenated** as the placenta passes oxygen into the fetal blood. This travels in the umbilical vein to the vena cava.
• Deoxygenated blood passes from the body into the right atrium of the heart.	• Oxygenated blood passes from the body into the right atrium of the heart.
	• **Some blood** travels across to the left atrium through a hole: the **foramen ovale**.
• Then it passes into the right ventricle.	• **A little blood** travels into the right ventricle.
• **All the blood** leaves the heart and travels to the lungs in the pulmonary artery.	• **A little blood** leaves the heart and travels towards the lungs in the pulmonary vein, but most crosses to the aorta through a connecting vessel: the **ductus arteriosus**.
• **Oxygenated** blood travels from the lungs to the left atrium and then the left ventricle of the heart.	• **Very little blood** travels from the lungs to the left atrium where it meets the blood from the right atrium.
• The left ventricle forces blood into the aorta.	• The left ventricle forces blood into the aorta where it meets the blood from the ductus arteriosus.
• **All the blood** travels to body organs and gives up its oxygen.	• **All the blood** travels to body organs and gives up its oxygen.

> **MUST REMEMBER**
> Very little blood goes to the lungs as the fetus does not use the lungs to oxygenate the blood.

CHANGES IN THE MOTHER DURING PREGNANCY

Factor	Effect

Thermal balance	• The fetus has a high metabolic rate.
	• Respiration in the fetus produces heat:
	  – Heat is passed to the mother's blood.
	  – The mother loses this extra heat.

Blood volume	• There is a high demand from the fetus for nutrients and oxygen.
	• The placenta grows and the volume of blood in the placenta increases:
	  – More plasma is produced.
	  – More red blood cells are produced.

Cardiac output	• The mother's muscles have to work harder due to her extra weight and the weight of the fetus.
	    Cardiac output = stroke volume × heart rate
	• Cardiac output increases by about 40%:
	  – The heart beats faster.
	  – There is a slight increase in stroke volume.

OXYGEN USED BY DIFFERENT STRUCTURES DURING PREGNANCY

Increased demand for oxygen	Little change in demand for oxygen
These structures increase in size during pregnancy: • Fetus • Placenta • Uterus • Breasts	Theses structures do not get much bigger during pregnancy: • Kidney • Lungs • Heart

SENESCENCE (AGEING)

Senescence is a continuous process starting at birth and ending at death. The effects of ageing become more obvious after 40 years.

CAUSES OF SENESCENCE

1. Mistakes in protein synthesis lead to chemical changes.
   - **Collagen** molecules in skin join together, making skin dryer.
   - **Elastin** molecules in skin become less elastic, making skin wrinkled.

2. There is a reduction in the rate of cell division.
   - Although all cells are capable of dividing, some lose this ability at birth:
     - Nerve cells are not replaced.
   - Cells that do retain their ability to divide do so at a slower rate as we age.

3. There is a decline in function of cells and organs.
   - This makes the body slower to respond to internal or external stimuli.

AGE CAUSES THE DECLINE OF SOME BODY FUNCTIONS

Body function	Explanation

BMR decreases	- BMR is a measure of the energy needed to keep the body functioning when at rest.
	- It is measured as the amount of energy used per kg of body mass per hour.
	- As the number of cells fall with age,
	  - **fewer active cells** means:
	  - a lower BMR.

Cardiac output decreases	- Muscle cells become **smaller** if not used.
	- Cells die as the heart ages.
	- This results is loss of cardiac strength and
	  - a **decrease** in the **stroke volume** of the ventricles.
	    Cardiac output = stroke volume × heart rate
	- So if stroke volume decreases then cardiac output falls.

Nerve conduction velocity decreases	- The rate at which nerve impulse are conducted is slower.
	- As the **myelin sheath** gets **thinner**:
	  - **ions** leak from the **axon**.
	- **Synaptic transmission** decreases:
	  - **less neurotransmitter** is produced at the synapse,
	  - so the threshold of the synapse is more difficult to overcome.
	All these result in a decrease in nerve conduction velocity.

Female reproductive function	**Menopause:** - Ovaries become insensitive to FSH. - No follicles will develop. - **No oestrogen** will be produced. - No ovulation occurs. - Female becomes **infertile**. Low oestrogen may result in: - increasing incidence of heart disease. - increasing loss of bone tissue – **osteoporosis**. Thus some women take **HRT** (hormone replacement therapy) to replace the oestrogen and so reduce the risk of heart disease and osteoporosis.

INDEX

absorption 69, 115
accommodation 46
Acquired Immunodeficiency
  Syndrome 99
actin 86, 128
action potential 41–2
active immunity 104
activity (exercise)
    and diet 122
    and muscular action 130
ADH (anti-diuretic hormone) 82
aerobic respiration 29–33, 34, 61
ageing (senescence) 131
AIDS 99
air, and transpiration 51
allele
  definition of 3
  frequencies of 12–13
allopatric speciation 16
amino acids
  and diet 119, 122
  and digestion 69, 77, 115
ammonia 77
anaerobic respiration 34
Animalia 18
animals, behaviour in 76
antibiotics 97
antibodies 97, 101, 104
antigens 97, 101
  non-self antigens 102, 103
aseptic techniques 91
ATP 29, 33, 129, 130
autonomic nervous system 76

bacterial cells 89–95
barriers to disease entry 100–1
Basal Metabolic Rate (BMR) 121, 131
batch culture 95
behaviour 75–6
bile 68, 114
binary fission 90
biotechnology, commercial 95
blood
  of the fetus 107
  fetus-placenta system 109–10
  supply in liver and kidney 53, 80
  transport of gases in 63–6, 123–6
  vessels, and digestion 69, 115
  water potential of 58, 82
blood glucose levels 37
blood volume, in pregnancy 110
BMR (Basal Metabolic Rate) 121, 131
body temperature 38

Bohr shift 64, 124
Bowman's (or renal) capsule 55, 56, 78, 79
brain 83–4
buccal cavity 68, 72, 114, 118

carbamino-haemoglobin 66, 126
carbohydrates 71, 119
carbon cycle 24
carbon dioxide 66, 126
cardiac muscle 85, 127
cardiac output
    in ageing 131
    in pregnancy 110
cells
    bacterial 89–95
    division of 1–2, 131
cellular immunity 102
cerebellum 83
cerebral cortex (or grey matter) 84
cerebrum 83, 84
chemoautotrophs 91
chemosynthesis 91
chi-squared test 8–10
chromosomes 1–2
    definition of 3
classification 17–18
codominant alleles, definition of 3
collagen 131
collecting duct 81
colour vision 48
commercial biotechnology 95
community, definition of 19
concentration 55, 57, 79, 81, 82
cone cells 48
continuous culture 95
convoluted tubules 55, 57, 79, 81
cornea 45
corpus luteum 106, 108
cortex, of root 49
counting cells 92–3
crosses, genetic 4–5
crossing over 2
culturing bacteria 91, 95

deamination 53, 77
deforestation 26
denitrification 25
diet 119–22
    and insect metamorphosis 73–4
diffusion 59, 81
digestion 67–72, 113–18
digestive enzymes 68, 74, 114
dihybrid dominant/recessive cross 7
dilution plating 93
disease, and microbes 89–104

disease-causing bacteria 98
disease-causing viruses 99–100
distal convoluted tubule 78, 79, 91
    see also second convoluted tubule
diversity 20–1
DNA, definition of 3
dominant allele, definition of 3
dominant/recessive cross
  dihybrid 7
  monohybrid 4
downstream processing 96
dry conditions, plant adaptation to 52
duodenum 68, 113, 114

ecology 19–27
effector 39
elastin 131
endodermis, of root 49
energy 120–1, 130
energy transfer 22–3
environment, definition of 19
environmental factors, and variation 11
enzymes
  digestive 68, 74, 118
  from micro-organisms 96
epidermis, of root 49
evolution 15
extracellular enzymes 96
eye 45–8

fallopian tube 105, 106, 107
fast twitch muscle fibres 130
fatty acids 69, 115
female
  growth in puberty 112
  reproductive system 105, 106, 131
fertilisation 107
fetal haemoglobin 65, 125
fetus-placenta blood system 65, 109–10, 125
fibres, muscle 85, 127
Fick's law 59
filtrate 56
filtration see ultrafiltration
first convoluted tubule 55, 57
    see also proximal convoluted tubule
fish 60, 62
five-kingdom classification 18
focus 46
follicular stage 106
food, and energy 120

forebrain 83
forests, and deforestation 26
fovea 45
Fungi 18

gall bladder 118
gametes 1, 105
gas exchange 59–62
gases, transport of 63–6, 123–6
gene, definition of 3
genetic crosses 4–5
genetics 3–10
genotypes 4, 5, 7
   definition of 3
   and variation 11, 13
gills 60
glucose
   absorption of 69, 115
   levels in blood 37, 53
glycerol 69, 115
glycogen 53, 122
glycolysis 29, 30–1
growth
   of bacterial cells 89–95
   measuring 111–12
gut wall 67–8, 113

habitat, definition of 19
haemocytometer 92
haemoglobin 63, 65, 123, 125
Hardy-Weinberg principle 12–13
hCG (human chorionic gonadotrophin) 108
heart, fetus-placenta blood system 110
herbivores 71
heterozygous genotype, definition of 3
hindbrain 83
HIV (Human Immunodeficiency Virus) 99
homeostasis 35–8
homeostatic organs, the liver 53
homologous chromosomes 1
homozygous genotype, definition of 3
homozygous recessive phenotype 14
hormones
   control of digestion 72, 118
   control of lactation 109
   and growth in puberty 112
   produced by placenta 108
host cell 98
human chorionic gonadotrophin (hCG) 108
Human Immunodeficiency Virus (HIV) 99

humidity, and transpiration 51
humoral immunity 103
hydrogen/electron transport chain 29, 33
hydrogencarbonate ions 66, 126
hypothalamus 82, 83

ileum 68, 70, 113, 114, 116
immobilised enzymes 96
immunity 103–5
implantation 107
independent assortment 2
infectivity 98
influenza virus 100
insects
   gas exchange in 60, 62
   metamorphosis and diet 73–4
intestine, small 70, 116
intracellular enzymes 96
invasiveness 98
involuntary (or unstriped) muscle 85, 127
iris 45
iron 119

kidney 55, 78–9
kidney (or renal) tubule (nephron) 55, 56, 78–9
kinesis 76
Krebs cycle 29, 32

lactation 109
lactose intolerance 117
larvae 73–4
leaves, spongy mesophyll 61
lens 45
life cycle, of insects 73–4
light dependence/independence 27–8, 51, 91
limiting factors 95
Lincoln Index 19
link reaction 29, 31
lipids 68, 114, 119
liver 53–4
loop of Henle 57, 81
lungs see respiration
luteal stage 106
lymph vessel 69, 115

male
   growth in puberty 112
   reproductive system 105–6
mammals 62, 71
maternal haemoglobin 65, 125
measuring growth 111–12
medulla 83
meiosis 1–2
menopause 131
menstruation 122
metabolism 121

metabolites, primary and secondary 97
metamorphosis 73–4
microbes, and disease 89–104
midbrain 83
milk (or lactose) intolerance 117
milk production (lactation) 109
minerals 119
monohybrid crosses 4–5
motor neurones 39
mucosa 67, 113
multiple alleles, definition of 3
muscle, of gut wall 67–8, 113
muscles 85–8, 127–30
myofibrils 85–6, 127–8
myoglobin 65, 125
myosin 86, 128, 129

nephron (kidney or renal tubule) 55, 56, 78–9
nerve conduction velocity 131
nervous control, of digestion 72, 118
nervous system 39–44, 76
nervous transmission 40–2
neurones 75
niche, definition of 19
nitrogen cycle 25–6
nitrogenous waste 54
non-self antigens 102, 103
non-specific response 100, 101
nutrients 119
nutrition, bacterial 91

oesophagus 67, 113
oestrogen 108, 109, 112, 131
oogenesis 106
ornithine cycle 77
osmoreceptors 82
ova and ovary 106–7
ovulation 106
oxidation 33
oxidative phosphorylation 33
oxygen 63–5, 123–4
   in pregnancy 110
oxygen dissociation curve 63–4, 65, 123–4, 125
oxyhaemoglobin 63, 123
oxytocin 108, 109

pancreas 118
parasites, viruses as 99
parasympathetic system 76
passive immunity 104
pathogens 98
penicillin 97
pH balance, effect of changes in 64, 124
phagocytosis 101

phenotypes 4, 5, 7
  definition of 3
  and variation 11, 14
phosphorylation 32, 33
photoautotrophs 91
photosynthesis 27–8, 61, 91
pituitary gland 82, 83
placenta 108
  fetus-placenta blood system 65, 109–10, 125
Plantae 18
plants
  gas exchange in 61, 62
  water transport in 49–52
plasma 66
population, definition of 19
pregnancy 110, 122
probability 8, 9, 10
progesterone 108, 109
Prokaryotae 18
prolactin 109
protein digestion 68, 114
proteins 71, 119
Protoctista 18, 59, 62
proximal convoluted tubule 78, 79, 81
  see also first convoluted tubule
pupil 45

rainforests 26
reabsorption 55, 57, 58, 79, 81–2
receptor 39
recessive allele, definition of 3
recessive cross see
  dominant/recessive cross
red blood cells 66, 126
reflex actions 72, 76
reflex arc 75
refraction 45
renal (or Bowman's) capsule 55, 56, 78, 79
renal (or kidney) tubule (nephron) 55, 56, 78–9
reproduction
  of bacterial cells 90
  male and female 105–10, 131

respiration 29–34, 66, 126
respiratory substrates 34
respiratory surfaces 59
response 39, 75, 76
  specific/non-specific 100, 101, 102
resting potential 40
retina 45, 46
rhodopsin 47
rod cells 47
roots 49–50
ruminant digestion 71

saltatory conduction 41
salts, absorption of 69, 115
sampling 19–20
sarcomeres 86, 87, 88, 129, 130
second convoluted tubule 55, 57
  see also distal convoluted tubule
selection 15–16
senescence (ageing) 131
sex cells (gametes) 1, 105
sex linkage 6–7
skeletal (or striped) muscle 85–6, 127
sliding filament theory 87, 129
slow twitch muscle fibres 130
small intestine 70, 116
speciation 15, 16
species, definition of 16
specific response 100, 102
speech areas of brain 84
sperm 105–6, 107
spinal cord and reflexes 75
spongy mesophyll 61
starch digestion 68, 114
statistical tests, chi-squared test 8–10
stimuli 39, 75, 76
stomach 67, 72, 113, 114, 118
stomata 51
submucosa 67, 68, 113
succession 21–2
sympathetic system 76

sympatric speciation 16
synapse 42–4

taxis 76
temperature
  of the body 38
  and transpiration 51
testosterone 112
thermal balance, in pregnancy 110
toxins 98
tracheal system 60
transmission, nervous 40–2
transpiration 50, 51
transport
  of gases 63–6, 123–6
  of water in plants 49–52
triglycerides 69, 115
trophic levels 22–3
tropomyosin 129
turbidimetry 92

ultrafiltration 55, 56, 79, 80
urea 53, 54, 77, 78
urine 54

vaccination 104
variation 11–14
vasodilation 38
vegetarian diet 122
viruses 99–100
visual association area of brain 84
vitamins 119

water, absorption of 69, 115
water homeostasis 58
water loss, limiting in gas exchange 62
water potential, of blood 58, 82
water reabsorption 82
water transport, in plants 49–52
weight loss, and diet 122

xerophytes 52
xylem 49–50